KEY TO THE FAMILIES OF FLOWERING PLANTS OF THE WORLD

£1.50

KEY TO THE FAMILIES OF FLOWERING PLANTS OF THE WORLD

REVISED AND ENLARGED
FOR USE AS A SUPPLEMENT TO

THE GENERA OF
FLOWERING PLANTS

J. HUTCHINSON
LL.D. (St. Andrews) F.R.S.

CLARENDON PRESS · OXFORD

Oxford University Press, Ely House, London W.1

GLASGOW NEW YORK TORONTO MELBOURNE WELLINGTON
CAPE TOWN SALISBURY IBADAN NAIROBI LUSAKA ADDIS ABABA
BOMBAY CALCUTTA MADRAS KARACHI LAHORE DACCA
KUALA LUMPUR HONG KONG TOKYO

First published 1967
Reprinted (with corrections) 1968

SET BY SANTYPE LIMITED OF SALISBURY
AND PRINTED IN GREAT BRITAIN
AT THE UNIVERSITY PRESS, OXFORD
BY VIVIAN RIDLER
PRINTER TO THE UNIVERSITY

CONTENTS

INTRODUCTION

THIS new edition of my *Key to the Families of Flowering Plants* has been revised and added to after further research and practical use over many years. It is here issued as a necessary supplement for use with my larger work, *The Genera of Flowering Plants*, of which Volume I was published on 4 December 1964 and Volume II on 11 May 1967. By its aid, following careful observation and dissection of floral parts, it is confidently expected that the *family* of a very large percentage of flowering plants, wild and cultivated in any part of the world, may be ascertained. In order to assist those with only a slight knowledge of botany, a short glossary has been added, together with a few diagramatic illustrations.

Compared with the original key, which appeared in the first and second edition of my *Families*, the main groups of the Dicotyledons have been subdivided into smaller units to facilitate more ready identification, mainly by separating those with *alternate* leaves (usually a more primitive character) from those with *opposite* leaves (more advanced). This will save the student from having to turn over more than a few pages to compare the contrasting key characters. As the key is mainly artificial, it frequently brings next to each other strange 'bedfellows', families which are quite unrelated, though sharing a very similar combination of characters, mostly due to convergent evolution.

First of all, to use the key effectively, it is essential that the gynoecium (pistil or ovary) should be carefully examined in order to ascertain whether it is composed of *free carpels* or whether these are united to form a *single ovary*. If the latter, then it is a simple operation to cut a cross-section with a sharp penknife, discarded razor-blade, or scalpel to find out if the ovules are inserted on the outer walls or in the middle, top or bottom of the ovary. If on the outer walls, one may turn at once to the 'SYNCARPAE' and its subdivision **'Parietales'**, in groups 5–11. If the ovules are in the middle or at the top or bottom of the ovary, then the specimen belongs to **'Axiles'**, in the remainder of the groups from 12 to 32.

Even some quite small and relatively homogeneous families require several entries in the key. For example in the holly family, *Aquifoliaceae*, composed of only four genera, the petals in some species are quite free (polypetalous), in others shortly united into a tube (sympetalous); in some the flowers are unisexual, in others bisexual, while the petals may be either imbricate or valvate, and the leaves may or may not have stipules.

Even in a single genus such as *Saurauia*, the petals may be free or partly united, and in some species there are no petals; so that we have here in one and the same genus flowers representing the main characters of the old groups *Polypetalae*, *Gamopetalae*, and *Apetalae*, of the de Candolle–Bentham & Hooker system.

In less homogeneous (less 'natural') families a greater number of different combinations of characters may be present, and it has been necessary to insert them in a large number of places in the key. For example the families *Flacourtiaceae* and *Euphorbiaceae* occur several times over. This should not be a reason for splitting up such familes into smaller units (except into subfamilies, tribes, and subtribes), for there are usually one or two characters which distinguish them to the experienced botanist. In *Flacourtiaceae* the placentation of the ovules is constantly *parietal* (a character found in several other related families); and in *Euphorbiaceae* the flowers are always *unisexual* and mostly reduced and accompanied by only 1 or 2 pendulous axile ovules in the ovary, and often with characteristic capsules and seeds, though with a great diversity in inflorescence, floral structure, fruits, and anatomy. The families are those that are recognized in my book, *The Families of Flowering Plants*, with a few additions which are marked by an *. These will be fully described in another work.

With practice the student may soon find that he can take a few short cuts to some of the groups in the key. For example if the stamens are found to be the same number as and opposite to the petals or corolla-lobes, the family he is looking for will be found in either groups 1, 17, 19, 23, 28, or 30. If the stamens are united into separate bundles he may turn at once to groups 6, 19, and 28. If the flowers are zygomorphic ('irregular'), the family will be included in groups 1 (a few *Helleboraceae*), 5, 6, 18, 24, 30, and 31. But this practice is not to be recommended, especially for beginners.

Do not, in using the key, read only the paragraph which seems to agree with the plant you wish to determine. You will be much more certain if you consult also the contrasting characters further on, which almost invariably begin with the same word or words. For example you may come to *Leaves gland-dotted*, and further on find *Leaves not gland-dotted*, inset at the same level, often on the same page, or if more distant at the page indicated in brackets.

Finally do not always blame the key if you fail to ascertain the family to which the plant belongs. It may be at fault, but you may have misread it or your own observations may be inaccurate. It is hoped, however, that by means of this key it should be possible to determine the family of most of the flowering plants to be met with on a day's march in any part of the world, from the north and south poles to the equator or from Greenland's icy mountains to India's coral strand. Some families, however, contain so many exceptions to the general characters that it is doubtful if the botanist has yet been born who can provide a key to the family of all *species* of flowering plants.

The plant illustrated on the cover is the "Oxford Ragwort", *Senecio squalidus* L., an alien from South Europe, first recorded on walls in Oxford in 1794 and since widely spread in Britain.

J. H., Kew, 1967.

KEY TO THE TWO PRINCIPAL GROUPS OF FLOWERING PLANTS

LEAVES typically net-veined, alternate or opposite, sometimes with stipules; flowers usually pentamerous or tetramerous; embryonic plant with 2 seed-leaves (cotyledons); vascular bundles of the stem usually arranged in a circle or circles (except in a few genera of the more primitive herbaceous families which have scattered bundles) DICOTYLEDONS (p. 3)

Leaves nearly always with parallel nerves, alternate, without stipules; flowers usually trimerous; embryonic plant with only 1 seed-leaf (cotyledon); vascular bundles of the stem closed and scattered (not in regular circles) MONOCOTYLEDONS (p. 101)

KEY TO THE ARTIFICIAL GROUPS OF DICOTYLEDONS[1]

Gynoecium (pistil) composed of 2 or more separate or nearly quite separate carpels with separate styles and stigmas (rarely the free carpels immersed in an expanded torus or tubular calyx as in some *Nymphaeaceae*, *Eupomatiaceae*, and *Rosa*) or enclosed by a large fleshy disk (*Capusiaceae*, *Scyphostegiaceae*) 'APOCARPAE'

Petals present, free from each other, sometimes considerably modified (as in some *Helleboraceae*):

Leaves alternate or all radical (sometimes peltate if in water) Group 1 (p. 6)

Leaves opposite or verticillate Group 2 (p. 11)

Petals present, more or less united into a tube (sympetalous) Group 3 (p. 13)

Petals absent, though sepals sometimes petaloid Group 4 (p. 14)

Gynoecium (pistil) composed of 1 carpel or of 2 or more united carpels with free or united styles, or if carpels free below (secondary apocarpy) then the styles or stigmas more or less united 'SYNCARPAE'

Ovules attached to the outer wall or walls of the ovary or to the partly intrusive placentas, if only 1 carpel then attached to the adaxial suture **'Parietales'**

Ovary superior:

Petals present, free from each other:

Leaves alternate or all radical, or reduced to scales Group 5 (p. 17)

Leaves opposite or verticillate Group 6 (p. 23)

Petals present, more or less united into a tube Group 7 (p. 25)

Petals absent Group 8 (p. 28)

Ovary inferior or semi-inferior

Petals present, free from each other Group 9 (p. 30)

Petals present, more or less united into a tube Group 10 (p. 32)

Petals absent Group 11 (p. 33)

Ovules attached to the central axis or to or near the base or apex of the ovary, sometimes only 1 ovule, sometimes the ovule or ovules pendulous from top of a basal placenta in a 1-locular ovary **'Axiles'**

Ovary superior, rarely partly (or at length wholly) immersed in the disk or calyx:

Petals present, free from each other or connate only at the base (very rarely connate in pairs but not into a complete tube):

Stamens more or less free from each other or connate only at or near the base, not united into separate bundles, sometimes united into 1 bundle (monadelphous):

[1] The names used for these groups are applied here for convenience. The 'Parietales' in this sense are not the *Parietales* of either Bentham & Hooker or of Engler & Prantl. The student should on no account mistake this artificial grouping for the system of classification in my *Families of Flowering Plants* (2nd edn. 1959, Clarendon Press, Oxford).

Stamens numerous, more than twice the number of the sepals or petals:
Leaves alternate or all radical — Group 12 (p. 34)
Leaves opposite or verticillate — Group 13 (p. 39)
Stamens not more than twice the number of the sepals or petals:
Flowers actinomorphic ('regular'):
Perfect stamens the same number as the petals and alternate with them, or up to twice as many, rarely fewer, rarely united into a column around the style:
Leaves simple (not unifoliolate) though sometimes much divided:
Leaves alternate or all radical — Group 14 (p. 41)
Leaves opposite (at least the lower ones) or verticillate — Group 15 (p. 48)
Leaves compound or unifoliolate — Group 16 (p. 52)
Perfect stamens the same number as (rarely fewer than) the petals and opposite to them (very rarely only 1) — Group 17 (p. 55)
Flowers zygomorphic ('irregular') — Group 18 (p. 57)
Stamens united into more than 1 separate bundles — Group 19 (p. 58)
Petals present, all of them more or less equally united into a lobed tube or cup:
Corolla actinomorphic ('regular') or nearly so:
Stamens when the same number as the corolla-lobes alternate with them, sometimes more numerous, rarely fewer:
Stamens more than twice as many as the corolla-lobes — Group 20 (p. 59)
Stamens as many as, fewer or up to twice as many as the corolla-lobes:
Leaves alternate or all radical, sometimes reduced to scales, or rarely absent — Group 21 (p. 60)
Leaves opposite or verticillate — Group 22 (p. 66)
Stamens the same number as the corolla-lobes and opposite to them (rarely fewer) — Group 23 (p. 69)
Corolla distinctly zygomorphic ('irregular') — Group 24 (p. 71)
Petals (and rarely also the calyx) absent, sometimes represented by scales:
Bisexual or male flowers (and often the female) with a distinct calyx (sometimes this petaloid):
Leaves alternate or all radical — Group 25 (p. 74)
Leaves opposite — Group 26 (p. 80)
Bisexual or male flowers (and mostly also the female) without a distinct calyx — Group 27 (p. 84)
Ovary more or less inferior or rarely semi-inferior:
Petals present, free from each other:
Leaves alternate, spirally arranged or all radical — Group 28 (p. 86)
Leaves opposite or verticillate, rarely reduced to scales — Group 29 (p. 90)

Petals present, more or less united (sympetalous) into a cup or tube:
  Leaves alternate, spirally arranged, or all or some radical — Group 30 (p. 93)
  Leaves opposite or verticillate — Group 31 (p. 96)
Petals absent — Group 32 (p. 98)

'APOCARPAE'. Group 1. *Gynoecium composed of* 2 *or more separate or nearly quite separate carpels with separate styles and stigmas; petals present, free from each other; leaves alternate or all radical.*

Leaves stipulate, sometimes the stipules minute or adnate to the petiole, or enclosing the young buds, rarely glandular:

Carpels numerous (rarely few and then stipitate), spirally arranged on a somewhat elongated floral axis or receptacle; sepals often 3 or indistinguishable from the 6 or more petals; flowers solitary, mostly large and conspicuous; seeds with copious endosperm and minute embryo; trees or shrubs *Magnoliaceae*

Carpels few or numerous on a small globose or concave floral axis or receptacle; sepals or calyx-lobes often 5; usually quite distinct from the petals:

Stamens free from one another or nearly so, or shortly united into separate bundles; calyx imbricate or valvate:

Carpels completely immersed in the greatly enlarged disk which covers them in fruit; stamens 5 *Capusiaceae*

Carpels not immersed in the disk as above:

Herbs more or less scapigerous with mostly radical sometimes compound leaves; seeds not arillate; endosperm usually copious; sepals mostly imbricate:

Petals and stamens hypogynous; stipules gland-like *Resedaceae*

Petals and stamens perigynous; stipules absent *Saxifragaceae*

Habit various; seeds not arillate; fruits follicular (dehiscent) or indehiscent, often the achenes arranged on or inside a large fleshy torus or receptacle; endosperm absent or very scanty; sepals mostly imbricate and partly united *Rosaceae*

Trees, shrubs, or climbers, very rarely herbs, with often scabrid prominently pinnately nerved leaves; stipules adnate to the petiole; seeds arillate, the aril often laciniate; sepals imbricate, often hardened in fruit, more or less free *Dilleniaceae*

Trees with large leaves and with stellate hairs; seeds not arillate; endosperm copious; calyx valvate *Tiliaceae*

Stamens more or less united into a column; calyx valvate; hairs on the leaves often stellate or lepidote *Sterculiaceae*

Leaves without stipules:

Carpels completely sunk in the tissue of the large broad torus:

Aquatic herbs with floating mostly peltate leaves and submerged rhizomes; endosperm of the seeds not ruminate *Nymphaeaceae*

Trees or shrubs; endosperm of the seeds ruminate; sepals and petals not differentiated from one another, forming a deciduous calyptra on the rim of the expanded concave torus *Eupomatiaceae*

Carpels not sunk in the torus or only slightly so (sometimes partly enclosed by the disk):

Stamens the same number as and opposite the petals; carpels usually 3, or rarely numerous (to p. 7):

Leaves compound; petals smaller than the sepals, sometimes minute; fruit usually baccate; endosperm never ruminate:

Carpels numerous, spirally arranged; leaves trifoliolate; flowers unisexual *Sargentodoxaceae*

Carpels few (up to 9) in whorls; leaves digitately compound or rarely pinnate or 3 foliolate *Lardizabalaceae*

Leaves simple or rarely trifoliolate or palmately lobed; fruit drupaceous; endosperm sometimes ruminate, the seeds often curved in the form of a horse shoe *Menispermaceae*

Stamens the same number and alternate with the petals or monadelphous, or more numerous (rarely fewer) than the petals:

Stamens numerous, more than 12, or at least more than double the number of the petals (to p. 9):

Sepals and petals connate into a calyptra:

Styles connate into a mass; endosperm ruminate; carpels sunk in the torus *Eupomatiaceae*

Styles free from each other; seeds not arillate; endosperm smooth; carpels sessile, not sunk in the torus:

Indumentum (hairs) not lepidote, often absent; leaves often gland-dotted; carpels remaining free in fruit *Winteraceae*

Indumentum lepidote; carpels concrescent in fruit; stamens with the connective prolonged far beyond the apex of the anther-loculi *Himantandraceae*

Styles free; seeds covered with a multifid aril; carpels on a short common stipe; leaves neither gland-dotted nor lepidote *Crossosomataceae*

Sepals and petals not connate into a calyptra; styles free or rarely slightly connivent at the base:

Sepals and petals in 3 distinct series (3+3+3) rarely in 2 series:

Anthers usually with a produced often broad and truncate connective; flowers usually bisexual; seeds with copious ruminate endosperm and minute embryo; trees, shrubs, or woody climbers; leaves alternate, entire *Annonaceae*

Anthers with a narrow and not produced connective; flowers dioecious *Menispermaceae*

Sepals and petals in 2 distinct series or rarely the sepals gradually passing into the petals, usually in fours or fives, or rarely the petals numerous; endosperm never ruminate:

Herbs, sometimes rather woody at the base (to p. 8):

Carpels in a single whorl and elevated on a stipe-like torus; petals 3- or more-times divided *Resedaceae*

Carpels usually spirally arranged, not elevated on the torus; petals entire, bifid, or reduced and tubular:

Carpels free from the beginning; sepals and petals usually 5 each, or the latter more numerous; flowers actinomorphic (regular) or zygomorphic (irregular); rarely aquatics; fruit an achene, follicle, or rarely a berry:

Disk present around the carpels, fleshy, sometimes nearly covering them; stamens centrifugal, numerous; seeds arillate; carpels very divergent in fruit; flowers large and showy *Paeoniaceae*

Disk absent; stamens centripetal; seeds not arillate:

Carpels with more than 1 ovule; fruits follicular or berry-like; flowers sometimes zygomorphic (irregular) *Helleboraceae*

Carpels with only 1 ovule; fruits a bunch of dry achenes, very rarely berry-like; flowers never zygomorphic (irregular) *Ranunculaceae*

Carpels free from the beginning; sepals and petals 3 each; flowers actinomorphic (regular); aquatic herbs with peltate floating leaves; ovules parietal, scattered within the carpel *Cabombaceae*

Carpels at first connivent, at length free and torulose; sepals 3; petals 6; flowers actinomorphic (regular); not aquatics *Papaveraceae*

Trees, shrubs, or woody climbers (from p. 7):

Petals and stamens hypogynous:

Flowers bisexual or rarely polygamous:

Calyx imbricate:

Leaves aromatic, pellucid-punctate; sepals deciduous; seeds not arillate; petals numerous *Illiciaceae*

Leaves not aromatic; lateral nerves prominent and parallel; sepals persistent and often accrescent; seeds mostly arillate *Dilleniaceae*

Leaves not aromatic; lateral nerves few; seeds arillate, aril multifid; sepals not accrescent *Crossosomataceae*

Calyx valvate:

Trees covered with peltate scales (lepidote); flowers with subulate staminodes; connective of anthers petaloid *Himantandraceae*

Trees and shrubs without scales but sometimes with stellate hairs; connective rarely petaloid:

Endosperm of seeds smooth; anther-connective not truncate at the apex; stamens several *Winteraceae*

Endosperm of seeds ruminate; anther-connective mostly truncate or produced beyond the loculi; stamens mostly numerous *Annonaceae*

Flowers unisexual:

Leaves simple; seeds with copious endosperm and small embryo *Schisandraceae*

Leaves pinnate or simple; seeds with or without endosperm and fairly large embryo *Simaroubaceae*

Petals and stamens perigynous:

Anthers opening by slits lengthwise; leaves not pellucid-dotted; disk usually absent:

Seeds arillate; endosperm present, fleshy; trees *Crossosomataceae*

Seeds not arillate; endosperm absent; some herbs as well as trees; flowers sometimes polygamo-dioecious *Rosaceae*

Anthers opening by valves; leaves mostly pellucid-dotted; disk present, adnate to the inside of the calyx-tube; calyx-lobes in 2 or more series; carpels on the inside of the hollow receptacle *Monimiaceae*

Stamens 12 or fewer, or double the number of the petals when more than 6 (from p. 7):
Leaves gland-dotted:
Leaves simple; carpels arranged in a spiral, mostly numerous; stamens more or less united in a mass; carpellary axis elongated in fruit; flowers solitary *Schisandraceae*
Leaves simple; carpels more or less in a single whorl; axis not elongated in fruit; flowers mostly cymose or fasciculate *Winteraceae*
Leaves often compound; carpels in a whorl; axis in fruit not elongated; stamens more or less free from one another *Rutaceae*
Leaves not gland-dotted:
Leaves compound:
Flowers unisexual or polygamous:
Carpels usually 3, rarely 6 or 9; stamens often monadelphous *Lardizabalaceae*
Carpels 5, nearly enclosed by the disk *Simaroubaceae*
Flowers bisexual:
Herbs; seeds with copious endosperm and small embryo, not arillate *Ranunculaceae*
Herbs, shrubs, or trees; seeds usually without endosperm:
Petals and stamens hypogynous:
Seeds not arillate; wood with resin ducts *Anacardiaceae*
Seeds mostly arillate; wood without resin ducts *Connaraceae*
Petals and stamens mostly distinctly perigynous; seeds not arillate; wood without resin ducts *Rosaceae*
Leaves simple (sometimes the submerged of aquatics much dissected):
Stamens free or slightly united only at the base (to p. 10):
Flowers bisexual (to p. 10):
Shrubs or trees; seeds usually arillate:
Style terminal or subterminal:
Leaves reduced to scales; aril of the seed laciniate; carpels all fertile *Dilleniaceae*
Leaves not reduced to scales; seeds not arillate; leaves pinnatisect; carpels 5-10, all fertile, with 2 ovules in the middle; stamens 5 or 10 *Helleboraceae*
Leaves not reduced; aril of seeds entire, more or less cupular; carpels all fertile with 2 basal collateral ovules *Connaraceae*
Leaves not reduced; seeds very small, not arillate; carpels 4, all fertile *Escalloniaceae*
Leaves not reduced and not pinnatisect; only 1 carpel fertile, the others reduced to styles; seeds without an aril; disk present *Anacardiaceae* (*Buchanania*)
Style lateral, almost at the base of each of the 5 carpels; seeds not arillate; leaves simple *Simaroubaceae*
Herbs; seeds not arillate:
Aquatic plants with peltate floating leaves *Cabombaceae*
Not aquatics:
Carpels at the top of a gynophore *Resedaceae*

Carpels not on a gynophore, but sometimes along an elongated torus:
Torus elongated or cone-like; annual herbs with entire leaves; carpels usually very numerous *Ranunculaceae*
Torus flat or concave; carpels few:
Petals more than 3:
Carpels the same number as the petals *Crassulaceae*
Carpels fewer than the petals *Saxifragaceae*
Petals 3; stamens 12; carpels united at the back with the calyx-tube *Aristolochiaceae*
Flowers unisexual, dioecious; petals and stamens in threes and multiples of 3; wood with broad medullary rays(from p. 9) *Menispermaceae*
Stamens united into a column; flowers unisexual:
Carpels numerous *Schisandraceae*
Carpels definite in number, often 3, rarely 6 or 9 *Menispermaceae*

'APOCARPAE'. Group 2. *Gynoecium composed of* 2 *or more separate or nearly quite separate carpels with separate styles and stigmas; petals present, free from each other; leaves opposite or verticillate (never all radical).*

Stamens numerous (15 or more):
Stamens arranged all to one side of the flower; seeds arillate; trailing or climbing shrubs *Dilleniaceae*
Stamens arranged symmetrically around the carpels:
Leaves stipulate; fruits follicular or indehiscent, sometimes inserted on a large fleshy torus; style often lateral or basal:
Stamens all fertile, not petaloid:
Seeds without endosperm; leaves not or rarely glandular-serrate; disk sometimes present and large, often lining the calyx-tube *Rosaceae*
Seeds with endosperm; leaves or leaflets often glandular-serrate *Cunoniaceae*
Stamens not all fertile, the outer petaloid, inner gradually smaller and sterile; petals gradually distinguished from the sepals *Austrobaileyaceae*
Leaves without stipules:
Herbs or succulent plants, sometimes slightly woody only at the base:
Carpels more or less free from the beginning; leaves often fleshy and connate at the base; flowers mostly 5-merous *Crassulaceae*
Carpels connivent at first, then free and torulose; lower leaves alternate, not fleshy; sepals 3; petals 6 *Papaveraceae*
Trees, shrubs or woody climbers:
Receptacle more or less campanulate or deeply concave or tubular; no tendrils:
Anthers opening by valves or slits; carpels arranged at or towards the base of a hollow receptacle; seeds with endosperm; mainly tropics *Monimiaceae*
Anthers opening by slits; sepals and petals in several series; carpels numerous, lining the hollow receptacle; seeds without endosperm; temperate regions *Calycanthaceae*
Receptacle neither hollow nor concave:
Petals numerous, linear; anther opening by longitudinal slits; petiole acting as a tendril; seeds not winged *Ranunculaceae*
Petals 5; shrubs with ovary of numerous nearly free carpels with winged seeds *Medusagynaceae*
Stamens less than 15 in number:
Herbs, often succulent; leaves mostly connate at the base, without stipules; flowers often cymose; carpels the same number as the petals; seeds minute, with fleshy endosperm *Crassulaceae*
Trees, shrubs, or woody climbers:
Leaves simple:
Carpels inserted at or near the base of a hollow receptacle (calyx-tube); anthers opening by valves or slits; seeds with endosperm *Monimiaceae*

Carpels inserted on a more or less convex or slightly concave torus:

Carpels 3–7; flowers large; petals yellow, not clawed, inserted at the base of a fleshy torus; stipules paired, axillary; anthers opening by short or long pores *Ochnaceae*

Carpels 5–10; petals persistent and thickened after flowering; shrubs with angular branches; stipules absent; flowers small, green; anthers large, opening by slits *Coriariaceae*

Carpels 1–3; petals often clawed, thin; stipules mostly intrapetiolar, often connate at the base; sepals often biglandular at the base; anthers opening by slits *Malpighiaceae*

Carpels 2; petals not clawed, scarious; flowers in axillary heads, sometimes unisexual; calyx not glandular; stipules caducous; anthers opening by slits *Cunoniaceae*

Leaves compound; leaflets mostly stipellate *Staphyleaceae*

'APOCARPAE'. Group 3. *Gynoecium composed of* 2 *or more separate or nearly quite separate carpels with separate styles and stigmas; petals present, more or less united into a tube* (*sympetalous*).

Stamens not inserted on the corolla-tube:
- **Leaves** simple; seeds sometimes with ruminate endosperm, not arillate:
  - **Flowers** bisexual; erect trees or shrubs with hard wood; sepals 3; corolla 6-lobed, lobes 2-seriate *Annonaceae*
  - **Flowers** dioecious; fruit a drupe; mostly climbers with soft wood and broad medullary rays *Menispermaceae*
- **Leaves** usually pinnate or unifoliolate; flowers bisexual; seeds often arillate; fruit dehiscent; trees, shrubs, or woody climbers:
  - **Leaves** not pellucid-punctate *Connaraceae*
  - **Leaves** pellucid-punctate *Rutaceae* (*Galipea*)
- **Leaves** simple, opposite or scattered, often fleshy, not punctate; flowers actinomorphic (regular), bisexual; fruit dehiscent; herbs or shrubs *Crassulaceae*

Stamens inserted on the corolla-tube; corolla-lobes contorted; leaves opposite *Apocynaceae*

'APOCARPAE'. Group 4. *Gynoecium composed of 2 or more separate or nearly quite separate carpels with separate styles and stigmas; petals absent.*

Trees, shrubs or hard-wooded climbers; leaves simple or rarely compound; sepals not or rarely slightly petaloid (to p. 15):
Leaves stipulate:
Stamens free or slightly connate only at the base:
Sepals or calyx-lobes imbricate:
Stipules free from or adnate to the petiole; fruits achenial or drupaceous, often enclosed by the tubular calyx; disk usually present and lining the calyx; flowers bisexual or polygamo-dioecious; leaves sometimes pinnate *Rosaceae*
Stipules laterally adnate to the petiole; fruits follicular; disk absent; flowers unisexual, dioecious; stamens 15–20, filaments long and filiform; carpels 4–6, with elongate filiform styles *Cercidiphyllaceae*
Stipules intrapetiolar, amplexicaul; fruits follicular; disk absent; flowers unisexual, polygamo-monoecious; stamens 6, in 2 whorls; Juan Fernandez *Lactoridaceae*
Sepals or calyx-lobes valvate:
Leaves alternate, simple, covered more or less with stellate hairs; disk absent; flowers bisexual *Sterculiaceae*
Leaves opposite or verticillate, usually compound; disk present; flowers dioecious; male flowers with a rudimentary ovary *Brunelliaceae*
Leaves opposite or verticillate, simple; male flowers when present without a rudimentary ovary; hypogynous scales sometimes present between the stamens and ovary *Cunoniaceae*
Stamens united or partially so into a column; anthers in a whorl or unequally arranged:
Sepals valvate; carpels not enclosed by a disk *Sterculiaceae*
Sepals imbricate; carpels numerous, enclosed by the enlarged fleshy globose disk (resembling an ovary); flowers dioecious, males in axillary and terminal racemes or lower flowers axillary; flowers within tiers of tubular truncate bracts *Scyphostegiaceae*
Leaves without stipules:
Leaves compound, not pellucid-punctate; stamens often 6 and monadelphous *Lardizabalaceae*
Leaves compound of 1-foliolate, pellucid-punctate *Rutaceae*
Leaves simple:
Seeds with ruminate endosperm; stamens numerous, mostly with a broad more or less truncate connective; carpels usually numerous, free in fruit *Annonaceae*
Seeds sometimes with ruminate endosperm; stamens definite in number, often in threes with a narrow connective; carpels few, free; mostly climbers *Menispermaceae*
Seeds with uniform endosperm; stamens usually with a very narrow

connective:

Stamens hypogynous or on a hypogynous disk:

Carpels in several series or in an irregular series, concrescent in fruit; trees or shrubs with lepidote indumentum *Himantandraceae*

Carpels mostly 3, free; woody climbers; indumentum not lepidote *Menispermaceae*

Carpels in a single series, free or nearly so in fruit; some herbs; indumentum not lepidote *Phytolaccaceae*

Stamens more or less perigynous or on a widened or hollow receptacle:

Receptacle obscure and solid, bearing the stamens on its outside; carpels in a single whorl, connate towards the base; anthers opening lengthwise by slits; ovules 1 to several in each carpel, pendulous *Trochodendraceae*

Receptacle more or less hollow, bearing the stamens on its inside; anthers often opening by valves *Monimiaceae*

Herbs with radical or alternate leaves, or very soft-wooded climbers with opposite leaves and broad medullary rays (from p. 14):

Carpels usually numerous, achenial and 1-seeded in fruit, often with long hairy tails, or follicular with several seeds; flowers mostly bisexual; sepals usually petaloid, valvate or imbricate:

Carpels with more than 1 ovule; fruits follicular, rarely baccate or connate and dehiscing at the apex *Helleboraceae*

Carpels with only 1 ovule; fruit a bunch of free dry (indehiscent) achenes, rarely berry-like *Ranunculaceae*

Carpels mostly few; sepals not petaloid, sometimes rather scarious or absent:

Flowers bisexual; stamens free amongst themselves; leaves alternate:

Carpels in a single whorl; sepals more or less free; style terminal or nearly so:

Leaves not modified into pitchers:

Stipules absent; flowers in terminal, leaf-opposed or axillary spikes of racemes; bracts small and not coloured or forming an involucre *Phytolaccaceae*

Stipules present, adnate to the petiole; flowers in dense spikes or racemes, often the upper leaves coloured and sometimes forming an involucre *Saururaceae*

Leaves modified into pitchers; flowers borne on a leafless raceme of cymules; calyx-lobes valvate; S.W. Australia *Cephalotaceae*

Carpels in more than 1 whorl; sepals free; style terminal or nearly so:

Carpels with more than 1 ovule; seeds with endosperm and small embryo; stipules absent *Helleboraceae*

Carpels with only 1 ovule; seeds with endosperm and small embryo; stipules absent *Ranunculaceae*

Carpels 1–4; sepals united into a tube; style usually basal or lateral; seeds without endosperm; stipules mostly adnate to the petiole *Rosaceae*

Carpels 1–4; sepals free; style absent, stigma terminal; small annual herb; stipules absent *Circaeasteraceae*

Flowers unisexual or polygamo-dioecious; sepals usually biseriately imbricate; stamens free or variously connate:

Climbers; carpels mostly 3 or 6; wood with broad medullary rays *Menispermaceae*

Herbs:

Fruiting carpels 5; stipules if present not adnate to the petiole; seeds with endosperm and curved embryo *Molluginaceae*

Fruiting carpels 1–3: (*Gisekia*)

Stipules present and partly adnate to the petiole; seeds without endosperm *Rosaceae*

Stipules absent; seeds with endosperm *Saxifragaceae*

Fruiting carpels 12–4, fleshy or drupe-like; stipules absent; embryo annular, surrounding the copious mealy endosperm *Phytolaccaceae*

'SYNCARPAE'. **'Parietales'.** Group 5. *Gynoecium composed of* 1 *carpel or of* 2 *or more united carpels with free or united styles, or if carpels free below, then the styles or stigmas united; ovules attached to the outer wall or walls of the ovary; ovary superior; petals present, free from each other; leaves alternate or all radical, or completely reduced* (as in most *Cactaceae*).

Fertile stamens numerous (more than 12) (to p. 19):
Filaments connate into a tube or column:
Anthers free, small; petals often valvate; flowers mostly in heads or dense spikes *Mimosaceae*
Anthers adnate to the tube, extrorse; glabrous aromatic trees; leaves pellucid-punctate; flowers cymose *Canellaceae*
Filaments connate into separate bundles:
Bundles of filaments opposite the sepals *Tiliaceae*
Bundles of filaments opposite the petals *Flacourtiaceae*
Filaments free or partly adnate to a gynophore, or shortly connate only at the base:
Ovary supported on a gynophore:
Fruits dehiscent, capsular; mostly herbs *Cleomaceae*
Fruits indehiscent; mostly trees and shrubs *Capparidaceae*
Ovary sessile or very nearly so, at most shortly stipitate:
Anthers horse-shoe-shaped; ovary-placentas 2; ovules numerous; stipules caducous, leaving a wide scar; leaves large; pedicels often with 5 large glands below the sepals *Bixaceae*
Anthers straight or nearly so; other characters not associated:
Anthers opening by apical pores or short pore-like slits:
Seeds hairy; ovary not lobed; leaves digitately nerved or lobed; fruit a capsule *Cochlospermaceae*
Seeds not hairy; ovary often deeply lobed, especially in fruit; leaves pinnately nerved; ovary often eccentric *Ochnaceae*
Anthers opening their full length by slits lengthwise:
Flowers actinomorphic ('regular') (to p. 18):
Fleshy plants often very spiny with fleshy leaves, or these completely reduced; no stipules; stigmas often many *Cactaceae*
Above characters not associated; leaves never completely reduced:
Stipules present, sometimes soon falling off, free or adnate to the petiole or glandular (to p. 18):
Sepals contorted in bud; petals very fugacious, often 4; placentas 3–5, parietal or adnate to the partially intrusive septa *Cistaceae*
Sepals imbricate or valvate; petals often 5:
Corona absent:
Ovary of more than 1 carpel, i.e. with 2 or more placentas:
Petals and stamens hypogynous:
Leaves mostly reduced: sepals persistent *Resedaceae*
Leaves simple; sepals at length reflexed or deciduous *Flacourtiaceae*

Leaves simple; sepals not reflexed *Tiliaceae*
Leaves mostly pinnate; sepals not or rarely reflexed, not deciduous *Anacardiaceae*
Petals and stamens perigynous:
Sepals persistent, accrescent *Flacourtiaceae*
Sepals not accrescent or only slightly so *Rosaceae*
Ovary of 1 carpel and with 1 placenta; stamens and petals hypogynous; sepals persistent, very imbricate *Dilleniaceae*
Corona present *Passifloraceae*
Stipules absent (from p. 17):
Trees, shrubs or woody climbers:
Leaves simple:
Flowers mostly solitary; endosperm of seeds ruminate; stamens and petals hypogynous *Annonaceae*
Flowers rarely solitary, sometimes spicate-racemose; endosperm smooth:
Ovary wholly superior:
Ovary of more than 1 carpel:
Leaves rather small or very small flowers often spicate or racemose *Tamaricaceae*
Leaves rather large; flowers rarely spicate *Flacourtiaceae*
Ovary of 1 carpel *Dilleniaceae*
Ovary semi-inferior *Flacourtiaceae*
Leaves mostly bipinnate; petals valvate; calyx tubular, often valvate; flowers mostly in heads or spikes *Mimosaceae*
Herbs:
Sepals more than 2; juice not milk-like:
Leaves ternately compound; sepals petaloid; carpel 1, baccate in fruit *Helleboraceae*
Leaves peltate, floating; carpels numerous; stigmas radiating and disk-like *Nymphaeaceae*
Leaves simple or variously divided; sepals not petaloid, 4; petals 4; ovary of 2 united carpels divided by a false septum *Brassicaceae* (*Cruciferae*)
Leaves simple, circinate in bud, usually very sticky-glandular; ovary not divided by a false septum *Droseraceae*
Sepals 2; juice milk-like; petals often 4, fugacious; ovary of more than 1 carpel; fruits sometimes linear *Papaveraceae*
Flowers zygomorphic ('irregular') (from p. 17):
Petals and stamens hypogynous; petals often considerably modified; ovary of 1 carpel *Helleboraceae*
Petals and stamens hypogynous or slightly perigynous; petals mostly laciniate; ovary of more than 1 carpel, sometimes gaping at the top *Resedaceae*
Petals and stamens perigynous or subperigynous; petals not lacinate:
Fleshy plants with thick stem and thick or entirely reduced leaves, mostly very spiny; ovary of more than 1 carpel *Cactaceae*
Not fleshy and leaves not reduced as above;
Ovary of 1 carpel:

Upper (adaxial) petal outermost; leaves simple, pinnate or unifoliolate, never bipinnate *Fabaceae* (*Papilionaceae*)
Upper (adaxial) petal innermost; leaves mostly pinnate or bipinnate, rarely unifoliolate *Caesalpiniaceae*
Ovary of 2 or more carpels; anthers opening by pores; leaves always simple *Polygalaceae*
Fertile stamens definite in number in relation to the sepals and petals, 12 or fewer, sometimes accompanied by several staminodes (from p. 17):
Stamens 6, four longer and two shorter (tetradynamous); sepals 4; petals 4; ovary of 2 carpels, mostly divided by a false septum *Brassicaceae* (*Cruciferae*)
Stamens not as above, if 6 then more or less of equal length; other characters not associated:
Flowers actinomorphic ('regular') or nearly so (to p. 21):
Parasitic nevergreen plants destitute of chlorophyll; anthers opening by slits lengthwise; ovules very numerous; seeds minute with undivided embryo *Monotropaceae*
Not parasitic; leaves more or less green; seeds with divided embryo:
Flowers with a distinct corona, this sometimes membranous or represented by a definite ring of hairs towards the base of the calyx-tube; ovary of more than 1 carpel:
Calyx-lobes (or sepals) imbricate:
Ovary 1-locular; fruit opening locucidally (if capsular); plants often with tendrils *Passifloraceae*
Ovary imperfectly 5-locular, with intrusive parietal placentas (but no central axis); capsule opening septicidally; leaves sheathing at the base *Greyiaceae*
Calyx-lobes valvate; styles widely separated at the base; calyx-tube long *Malesherbiaceae*
Flowers without a corona:
Stamens completely united into a column with the anthers extrorse on the outside; leaves pellucid-punctate; endosperm not ruminate *Canellaceae*
Stamens united into a column with the anthers in a ring around the apex; leaves not punctate; endosperm often ruminate *Menispermaceae*
Stamens united into a tube; leaves pinnate; disk present *Meliaceae*
Stamens free or united only at the base, or rarely only the anthers connivent:
Leaves with stipules; these free or rarely adnate to the petiole (to p. 20):
Anthers with the connective produced above the loculi:
Anthers connivent around the style or subsessile *Violaceae*
Anthers not connivent around the style *Flacourtiaceae*
Anther-connective not produced:
Staminodes present, sometimes petaloid; anthers usually opening by terminal pores; stipules sometimes pectinate or ciliate *Ochnaceae*
Staminodes absent:
Leaves with very numerous and very sticky gland-tipped hairs or processes, mostly circinate in bud; seeds not carunculate *Droseraceae*

Leaves without sticky hairs and not circinate in bud; seeds often carunculate or arillate:
Styles free to the base *Turneraceae*
Style single: *Saxifragaceae* (*Mitella*)
Ovary of more than 1 carpel:
Mostly climbers with tendrils *Passifloraceae*
Shrubs or trees without tendrils *Passifloraceae*
Herbs with mostly stipitate siliquiform capsules; leaves sometimes compound *Cleomaceae*
Ovary of 1 carpel; petals marcescent *Sterculiaceae* (*Waltheria*)
Leaves without stipules (from p.19):
Stamens usually the same number as and opposite to the petals:
Sepals and petals usually in threes; anthers sometimes opening by valves:
Leaves 2-3-times pinnately compound, with swollen joints; erect shrubs; sepals numerous, spirally arranged; stamens 6; anthers subsessile *Nandinaceae*
Leaves mostly all radical, peltate, palminerved or lobed or 2–3-foliolate; perennial herbs; carpel 1 *Podophyllaceae*
Leaves simple or pinnate, often with prickly margins; shrubs or shrublets with hard wood; carpel 1 *Berberidaceae*
Sepals and petals 2 and 4 respectively; flowers solitary or terminal or leaf-opposed; anthers opening by slits *Fumariaceae*
Stamens with at least one row alternate with the petals:
Herbs or subshrubs:
Calyx tubular; stamens usually perigynous; seeds arillate, pitted *Turneraceae*
Calyx-tube short or absent; stamens hypogynous:
Ovary and fruit not stipitate:
Leaves peltate; carpel 1 *Podophyllaceae*
Leaves not peltate, mostly glandular and not cordate *Droseraceae*
Leaves neither peltate nor glandular, often cordate; staminodes present, sometimes multifid *Parnassiaceae*
Ovary and fruit stipitate *Cleomaceae*
Calyx-lobes free to the base *Achariaceae*
Trees, shrubs or climbers:
Leaves simple, but sometimes deeply divided (to p. 21):
Leaves very small and more or less scale-like or ericoid; flowers very small, in slender catkin-like spikes or racemes; seeds hairy *Tamaricaceae*
Leaves and other characters not as above:
Leaves digitately lobed *Caricaceae*
Leaves pinnately lobed or nerved or subdigitately nerved; calyx-tube very short; stamens hypogynous:
Staminodes absent:
Disk absent; leaves not glandular-serrate:
Anthers versatile, more or less sagittate *Pittosporaceae*

Anthers basifixed, not sagittate *Philadelphaceae*
Disk present; leaves often glandular-serrate *Escalloniaceae*
Staminodes present, in a ring outside the stamens; placentas partly touching in the middle but not forming a central axis *Greyiaceae*
Leaves pinnately nerved or lobed; calyx-tube long; stamens mostly perigynous *Turneraceae*
Leaves compound (from p. 20):
Flowers in heads or dense spikes; petals valvate; ovary of 1 carpel; ovules more than 1 *Mimosaceae*
Flowers neither in heads nor dense spikes; ovary usually of more than 1 carpel; ovule 1 or 2 and collateral:
Ovule 1, suspended on a basal funicle *Anacardiaceae*
Ovules 2, collateral, ascending from the base of the inner angle *Connaraceae*
Flowers racemose; ovary of 5 carpels with numerous ovules *Caricaceae*

Flowers markedly zygomorphic ('irregular') (from p. 19):
Stamens 4 or 6, opposite the petals, more or less united into 2 bundles; inflorescence often leaf-opposed *Fumariaceae*
Stamens with at least one row alternate with the petals:
Fertile stamens about 1 or 2; ovary often supported on a short to long gynophore:
Ovary composed of more than 1 carpel, with 2 or more placentas:
Fruits dehiscent; replum present *Cleomaceae*
Fruits indehiscent; no replum *Capparidaceae*
Ovary composed of 1 carpel (1 placenta); leaves usually pinnate *Caesalpiniaceae*
Fertile stamens more than 2:
Fleshy plants with numerous petals and mostly spiny and without leaves *Cactaceae*
Not fleshy; petals few:
Herbs or rarely shrubs or trees; anthers often with produced connective, mostly connivent or connate around the style; lowermost petal often spurred *Violaceae*
Herbs, shrubs or trees; anther-connective not produced or only glandular at the apex:
Fruits gaping at the top before maturity; stipules gland-like *Resedaceae*
Fruits closed until mature; stipules if present not gland-like:
Placentas 3; 5 fertile stamens with the same number of staminodes; ovary shortly stipitate; disk lining the calyx-tube *Moringaceae*
Placentas 2; ovary stipitate; disk absent *Cleomaceae*
Placenta 1:
Upper (adaxial) petal innermost; petals not markedly different, though sometimes only 1; anthers not dimorphous *Caesalpiniaceae*
Upper (adaxial) petal outermost; petals usually very dissimilar,

composed of standard (vexillum), wings, and keel (papilionaceous) *Fabaceae* (*Papilionaceae*)

Petals imbricate; anthers opening by pores *Polygalaceae*

'SYNCARPAE'. **'Parietales'.** Group 6. *Gynoecium composed of* 1 *carpel or of* 2 *or more united carpels with free or united styles, or if carpels free below, then the styles or stigmas united; ovules attached to outer wall or walls of the ovary; ovary superior; petals present, free from each other; leaves opposite or verticillate.*

Stamens more or less united into three or more separate bundles; leaves often gland-dotted or streaked with resin-canals:
 Mostly herbs or shrublets; leaves often with pellucid dots or black glands; styles free or nearly so; cotyledons distinct and well developed in the seed *Hypericaceae*
 Trees or shrubs; leaves with lines or resin-canals and very numerous lateral nerves; stigma sessile or subsessile; cotyledons very small or obsolete in the seed *Clusiaceae* (*Guttiferae*)
Stamens free or more or less united at the base or higher into not more than 2 separate bundles (diadelphous), sometimes adnate to a gynophore; leaves not glandular or rarely so:
 Connective of the anthers produced above the loculi; flowers often somewhat zygomorphic ('irregular'), the lower petals often gibbous, spurred or saccate at the base; leaves stipulate *Violaceae*
 Connective of the anthers not produced; flowers usually actinomorphic ('regular'):
  Stamens 6, tetradynamous (4 longer and 2 shorter); sepals 4; petals 4, often clawed; ovary composed of 2 carpels mostly divided by a false septum *Brassicaceae* (*Cruciferae*)
  Stamens and other characters not as above, stamens often more numerous:
   Stamens more than double the number of the petals:
    Ovary stipitate:
     Stamens hypogynous, not inflexed in bud; calyx-lobes without accessory teeth *Capparidaceae*
     Stamens perigynous, inflexed in bud, inserted on the calyx-tube, the lobes of the latter with accessory teeth *Lythraceae*
    Ovary sessile:
     Petals and stamens hypogynous:
      Sepals imbricate:
       Style single or sessile stigmas 1–3; indumentum when present often stellate; leaves not gland-dotted; petals fugacious *Cistaceae*
       Styles or sessile stigmas 2 or more; leaves often gland-dotted or with black glands *Hypericaceae*
      Sepals induplicate-valvate; leaves mostly subconnate at the base, often small and ericoid *Frankeniaceae*
     Petals and stamens perigynous; leaves not gland-dotted:
      Anthers not inflexed in bud *Flacourtiaceae*
      Anthers inflexed in bud *Sonneratiaceae*
   Stamens the same as or double the number of the petals (rarely fewer):
    Stamens hypogynous to perigynous; corona present; stamens adnate to a gynophore *Passifloraceae*

Stamens and petals hypogynous; style one, divided into as many stigmas as placentas; no corona:

Leaves small and ericoid, often more or less connate at the base; flowers usually sessile; sepals induplicate-valvate; petals clawed, with a scale-like appendage at the base *Frankeniaceae*

Leaves not small and not connate at the base; flowers pedicellate *Flacourtiaceae*

Stamens and petals more or less perigynous; no corona:

Stamens diadelphous or monadelphous; flowers zygomorphic ('irregular'): *Fabaceae* (*Papilionaceae*)

Stamens not diadelphous; flowers actinomorphic ('regular'):

Styles free (more than 1) or united only at the base; seeds with endosperm; stamens erect in bud:

Herbs with fleshy leaves without stipules; carpels as many as petals *Crassulaceae*

Shrubs with deciduous bark and thin leaves; carpels fewer than the petals; no gynophore:

Indumentum mostly of stellate hairs *Philadelphaceae*

Indumentum not stellate or absent *Hydrangeaceae*

Mostly climbers with tendrils; leaves usually stipulate; ovary seated on a gynophore *Passifloraceae*

Styles more or less united or style 1; seeds without endosperm; stamens inflexed in bud, rarely reduced to 2 *Lythraceae*

'SYNCARPAE'. **'Parietales'.** Group 7. *Gynoecium composed of* 1 *carpel or of* 2 *or more united carpels with free or united styles, or if carpels free below, then the styles or stigmas united; ovules attached to the outer wall or walls of the ovary; ovary superior; petals present, more or less united into a tube (sympetalous).*

Stamens free from the corolla-tube (to p. 26):
Ovary composed of more than 1 carpel (to p. 26):
Stamens numerous, more than twice the number of the corolla-lobes:
Anthers opening by a longitudinal slit, with a broadened truncate connective; seeds with copious ruminate endosperm; shrubs, trees, or climbers with exstipulate simple entire leaves *Annonaceae*
Anthers without a broadened connective; fleshy plants with reduced leaves and often very spiny:
Stamens hypogynous, inserted below a disk; leaves often fasciculate; corolla fairly long *Fouquieriaceae*
Stamens perigynous; disk absent; reduced leaves or leaves absent; usually very spiny *Cactaceae*
Anthers without a broadened connective; plants not fleshy; leaves present and mostly coriaceous:
Flowers spicate-racemose; petals with induplicate margins *Flacourtiaceae*
Flowers cymose-paniculate, with subscorpioid branches; corollal-lobes overlapping, 14–11, in 3 or 4 irregular series *Hoplestigmataceae*
Stamens definite in number in relation to the corolla-lobes:
Anthers opening by terminal pores:
Leaves alternate, pinnately nerved; stamens 4, connate at the base or up to the middle *Polygalaceae*
Leaves mostly opposite, with longitudinal parallel main nerves between the midrib and margins *Melastomataceae*
Anthers opening by slits lengthwise; leaves alternate or fasciculate, without stipules:
Branches climbing by hooks; calyx-lobes becoming enlarged and wing-like in fruit; petals contorted *Ancistrocladaceae*
Branches not hooked, but sometimes tendrils present:
Leaves fairly large; flowers more or less corymbose or paniculate:
Plants not spiny, or very rarely so; leaves not fasciculate; sepals 5, imbricate in bud; stamens 5, petals 5, at most connivent *Pittosporaceae*
Spiny shrubs; leaves often fasciculate, small and fleshy; corolla-tube fairly long; Mexico and nearby United States *Fouquieriaceae*
Trees, not spiny; sepals minute or absent; petals 4, in a subglobose tube *Violaceae*
Herbs, sometimes climbing; sepals 2 *Fumariaceae*
Leaves very small and scale-like; flowers in slender spikes or racemes *Tamaricaceae*

Ovary composed of a single carpel (usually a legume); stamens either free or more usually diadelphous or monadelphous, often 10, rarely more numerous (from p. 25):

Stamens connate into a sheath or tube, or free; flowers mostly bisexual:

Flowers actinomorphic ('regular'); petals valvate; calyx gamosepalous or valvate; leaves usually bipinnate, rarely simply pinnate or reduced to phyllodes; flowers often collected into heads *Mimosaceae*

Flowers zygomorphic ('irregular') or rarely actinomorphic; sepals imbricate or rarely valvate; petals imbricate, the upper (adaxial) one inside the others; leaves often pinnate or bipinnate *Caesalpiniaceae*

Flowers zygomorphic ('irregular') petals imbricate, the upper (adaxial) one (the standard) outside the others, the lateral 2 (the wings) outside the abaxial pair (the keel) which are sometimes loosely united along their lower edges; leaves simple, digitate, simply pinnate or unifoliolate *Fabaceae* (*Papilionaceae*)

Stamens connate into a column with the anthers in a ring at the top; flowers dioecious; fruit a drupe; slender climbers *Menispermaceae*

Stamens inserted on the corolla-tube or sometimes clearly adnate to it near the base (from p. 25):

Stamens double the number of the corolla-lobes or more numerous:

Plants not spiny and not cactus-like:

Flowers zygomorphic ('irregular') *Polygalaceae*

Flowers actinomorphic ('regular') *Caricaceae*

Plants usually very spiny; cacti *Cactaceae*

Stamens the same number as the corolla-lobes; corolla actinomorphic or nearly so (to p. 27):

Leaves alternate or radical:

Corolla-lobes without opposite glands; flowers bisexual:

Corolla-lobes imbricate or rarely contorted *Hydrophyllaceae*

Corolla-lobes induplicate-valvate; leaves simple or trifoliolate, sometimes orbicular and peltate; aquatic or marsh herbs; anthers sagittate and versatile *Menyanthaceae*

Corolla-lobes with glands opposite to them; flowers unisexual, monoecious; calyx very short; South Africa *Achariaceae*

Leaves opposite or subverticillate:

Pollen granular, not in waxy masses; corona absent:

Ovules 4, parietal in pairs, the lower pair erect, the upper pair pendulous; seeds with a tuft of hairs at the apex; style twice lobed in the upper part *Plocospermaceae*

Ovules not arranged as above; seeds without a tuft of hairs:

Trees, shrubs, or woody climbers:

Stamens 5 or more; ovules numerous *Potaliaceae*

Stamens 4:

Ovules numerous; some climbers *Bignoniaceae*

Ovules 1 or 2 in each loculus *Verbenaceae*

Herbs:

Stamens inserted in the tube of the corolla; corolla-lobes contorted *Gentianaceae*

Stamens inserted near the base of the corolla; corolla-lobes imbricate *Hydrophyllaceae*

Pollen in waxy masses; fruit follicular, dehiscing ventrally; seeds with a tuft of hairs at the apex; stamens with a coronal appendage *Asclepiadaceae*

Stamens fewer than the corolla-lobes, 4 or 2; corolla zygomorphic ('irregular') or rarely subactinomorphic ('subregular') (from p. 26):

Leafless nevergreen herbs parasitic on the roots of other plants; seeds minute, very numerous, with endosperm *Orobanchaceae*

Leafy green plants, rarely parasitic:

Ovules numerous on each placenta:

Mostly trees or woody climbers; leaves mostly compound, often ending in tendrils; seeds transverse to the length of the fruit, winged *Bignoniaceae*

Mostly herbs; seeds minute, not winged:

Flowers usually with characteristic glands (metamorphosed flowers) at the base of the pedicels; anther-loculi more or less parallel; ovary superior or inferior; Old World *Pedaliaceae*

Flowers without glands as above; anther-loculi divaricate; ovary superior:

Fruits horned *Martyniaceae*

Fruits not horned *Gesneriaceae*

Ovules 1 or 2 on each placenta; stem and branches often 4-sided *Verbenaceae*

'SYNCARPAE'. **'Parietales'.** Group 8. *Gynoecium composed of* 1 *carpel or of* 2 *or more united carpels with free or united styles, or if carpels free below, then the styles or stigmas united; ovules attached to the outer wall or walls of the ovary; ovary superior; petals absent.*

Leaves modified into pitchers; flowers dioecious; stamens united into a column; ovary 3- or 4-locular with numerous ovules in each loculus; seeds elongate-fusiform; embryo straight *Nepenthaceae*
Leaves not modified into pitchers:
 Submerged aquatic moss- or alga-like herbs; flowers dioecious; ovules numerous in each loculus; seeds microscopic, compressed, devoid of endosperm *Hydrostachyaceae*
 Not submerged etc. as above:
  Stamens 6, 4 long and 2 short (tetradynamous); fruits often with a thin membranous false septum between the placentas; seeds without endosperm with accumbent or incumbent cotyledons; sepals usually 4 *Brassicaceae* (*Cruciferae*)
  Stamens rarely 6 and then not tetradynamous:
   Stamen 1:
    Stems not jointed; leaves well developed; flowers bisexual, spicate; anther erect in bud *Lacistemataceae*
    Stems jointed; leaves reduced to scales; flowers unisexual, males spicate; anther inflexed in bud *Casuarinaceae*
   Stamens more than 1:
    Ovary composed of 1 carpel:
     Stamens numerous:
      Flowers zygomorphic ('irregular'); fruit mostly a legume *Fabaceae* (*Papilionaceae*)
      Flowers actinomorphic ('regular'); fruit a berry *Flacourtiaceae*
     Stamens 10 or fewer only by abortion:
      Leaves usually compound; flowers zygomorphic ('irregular') *Caesalpiniaceae*
      Leaves mostly simple; flowers actinomorphic ('regular'):
       Leaves stipulate:
        Anthers erect in bud *Ulmaceae*
        Anthers inflexed in bud *Moraceae*
       Leaves without stipules:
        Calyx of the male flower none; female calyx minute, consisting of small unequal connate scales; ovule 1 *Leitneriaceae*
        Calyx of the male flower represented by one unilateral scale, absent from the female, but female flowers surrounded by several series of imbricate bracts; ovules 2, collateral *Balanopsidaceae*
     Stamens 6; trees with bitter bark *Simaroubaceae*
     Stamens 4, opposite the valvate, often petaloid calyx-segments which often resemble a corolla; mostly Southern Hemisphere *Proteaceae*
    Ovary composed of more than 1 carpel, at least with 2 or more placentas

or more than 1 ovule:
Ovary and fruit stipitate:
Flowers not in catkins:
Stamens more than 4; sepals rarely valvate, not or only slightly coloured; flowers not in heads; fruit indehiscent *Capparidaceae*
Stamens numerous; seeds arillate *Papaveraceae*
Stamens 4; sepals valvate, often coloured; flowers often in heads *Proteaceae*
Flowers in catkins, dioecious, often produced before the leaves; stamens 2 or more; seeds with numerous fine hairs *Salicaceae*
Ovary sessile:
Flowers with a distinct corona; sepals and stamens more or less perigynous *Passifloraceae*
Flowers without a corona but sometimes with a hypogynous disk:
Stamens hypogynous or flowers unisexual:
Filaments connate into a column, anthers up to 20, adnate to the outside of the tube; leaves pellucid-punctate *Canellaceae* (*Canella*)
Filaments free or slightly connate only at the base:
Leaves alternate or fasciculate or much reduced:
Trees or shrubs; stipules small, caducous or absent; seeds with endosperm *Flacourtiaceae*
Shrubs; stipules absent; seeds without endosperm *Passifloraceae*
Herbs with large pinnately or digitately nerved leaves; no stipules; flowers paniculate; seeds with endosperm and small embryo, sometimes arillate *Papaveraceae*
Spiny much-branched shrubs with slender virgate leafless branches; fruit a berry; no endosperm *Resedaceae*
Leaves opposite; mostly shrubs; stipules usually present:
Sepals contorted; stamens not petaloid; shrubs or herbs often with stellate indumentum *Cistaceae*
Sepals imbricate; stamens petaloid; large climbing shrubs *Austrobaileyaceae*
Stamens distinctly perigynous:
Anthers inflexed in bud; staminodes rarely present; filaments free *Lythraceae*
Anthers not inflexed in bud; staminodes often alternating with the fertile stamens; filaments free or connate; indumentum sometimes stellate *Flacourtiaceae*
Anthers not inflexed in bud; no staminodes; stamens 4, opposite the valvate (often petaloid) sepals *Proteaceae*

'SYNCARPAE'. **'Parietales'.** Group 9. *Gynoecium composed of* 1 *carpel or of* 2 *or more united carpels with free or united styles, or if carpels free below, then the styles or stigmas united; ovules attached to the outer wall or walls of the ovary; ovary inferior or semi-inferior; petals present, free from each other.*

Aquatic herbs with floating or submerged leaves; petals numerous in several series; fruit a several-locular berry *Nymphaeaceae*
Not aquatic, rarely marsh plants:
Flowers bisexual (sometimes outer flowers sterile) (to p. 31):
Anthers opening by terminal pores:
Leaves not gland-dotted, mostly opposite with longitudinal main nerves; stamens definite, often double the number of the petals *Melastomataceae*
Leaves gland-dotted, more or less pinnately nerved *Myrtaceae*
Anthers opening lengthwise by slits:
Ovary-loculi superposed in two series; leaves opposite or subopposite; fruit a spherical edible berry crowned by the calyx-limb; anthers dorsifixed *Punicaceae*
Ovary-loculi not superposed:
Stamens numerous, more than twice the number of the petals:
Shrubs or trees with gland-dotted mostly opposite or rarely alternate leaves; placentas often 2, sometimes sub-basal or excentric *Myrtaceae*
Leaves not gland-dotted, sometimes fleshy or much reduced or absent:
Fleshy herbs or shrubs, often very spiny and in semi-desert regions; leaves reduced or absent *Cactaceae*
Herbs with rough hairs or rarely woody; stamens often in bundles opposite the petals; inflorescence often leaf-opposed; style single *Loasaceae*
Herbs without rough hairs; stamens 15 or more, not in bundles *Hydrangeaceae*
Trees or shrubs:
Sepals gradually passing into the petals; endosperm ruminate *Eupomatiaceae*
Sepals distinct from the petals; endosperm not ruminate *Flacourtiaceae*
Stamens definite in number, 4–12:
Leaves alternate or fasciculate (to p. 31):
Shrubs or trees; calyx imbricate or valvate:
Leaves without stipules, often glandular:
Disk present; leaves often glandular-serrate *Escalloniaceae*
Disk absent; indumentum mostly of stellate hairs; filaments often toothed near the apex *Philadelphaceae*
Leaves mostly with stipules; stipules more or less adnate to the petiole; shrubs, often resinous-glandular and sometimes spiny *Grossulariaceae*

Herbs or climbers:
Petals contorted; ovary usually 4-locular *Onagraceae*
Petals imbricate or valvate:
Herbs or climbers often rough with stinging hairs *Loasaceae*
Herbs with perennial rhizomes, often glandular *Saxifragaceae*
Leaves opposite, without stipules (from p. 30):
Leaves pinnately nerved, not gland-dotted; outer flowers sometimes sterile and without petals *Hydrangeaceae*
Leaves with main nerves parallel with the margin, not gland-dotted; flowers all fertile *Melastomataceae*
Leaves pinnately nerved, gland-dotted *Myrtaceae*
Flowers unisexual (from p. 30):
Leaves gland-dotted; style simple *Myrtaceae*
Leaves not gland-dotted; style or stigmas usually 3:
Shrubs, often armed with spines; leaves plicate or convolute in bud *Grossulariaceae*
Herbs or herbaceous climbers:
Stipules present; no tendrils; stamens numerous; anthers all 2-locular; ovary and fruit often winged *Begoniaceae*
Stipules absent:
Stamens 3 or 5, free or united; one anther often 1-locular, tendrils often present *Cucurbitaceae*
Stamens 4-many; anthers all 2-locular; no tendrils *Datiscaceae*

'SYNCARPAE'. **'Parietales'.** Group 10. *Gynoecium composed of* 1 *carpel or of* 2 *or more united carpels with free or united styles, or if carpels free below, then the styles or stigmas united; ovules attached to the outer wall or walls of the ovary; ovary inferior; petals present, more or less united into a tube (sympetalous).*

Stamens numerous; leaves often gland-dotted, mostly opposite *Myrtaceae*
Stamens usually definite in number, rarely more than twice the number of the corolla-lobes:
Fleshy plants (often), usually with minute leaves or leafless; calyx-lobes, petals and stamens numerous; style with radiate stigmas *Cactaceae*
Above characters not combined; stamens mostly the same or double the number or fewer than the petals:
Flowers unisexual; leaves alternate; tendrils often present; stamens mostly 3, the anthers often twisted *Cucurbitaceae*
Flowers bisexual; leaves opposite or verticillate; no stipules; anthers mostly opening by terminal pores *Melastomataceae*
Flowers bisexual, very rarely unisexual: leaves opposite or alternate, often with stipules; anthers not opening by pores, sometimes connivent at the apex:
Leaves with interpetiolar or intrapetiolar stipules; flowers actinomorphic ('regular'); stamens usually the same number as the corolla-lobes:
Ovules numerous on the walls of the ovary; branches not hooked *Rubiaceae (Gardenia)*
Ovule solitary; branches hooked; calyx-lobes enlarged in fruit *Ancistrocladaceae*
Leaves without stipules (sometimes the leaves anisophyllous, the smaller resembling a stipule):
Stamens numerous:
Plants often roughly hairy; corolla more or less perigynous or epigynous; filaments free or collected into bundles opposite the petals *Loasaceae*
Plants not roughly hairy; corolla hypogynous *Flacourtiaceae*
Stamens 4 or 2:
Evergreen trees or shrubs; corolla subactinomorphic (slightly 'irregular'); stamens 2, with very broad connective and sinuous anther-loculi; leaves opposite *Columelliaceae*
Usually herbaceous, rarely subwoody plants; corolla more or less zygomorphic ('irregular'); stamens 4 or 2, often didynamous *Gesneriaceae*

'SYNCARPAE'. **'Parietales'.** Group 11. *Gynoecium composed of* 1 *carpel or of* 2 *or more united carpels with free or united styles, or if carpels free below, then the styles or stigmas united; ovules attached to the outer wall or walls of the ovary; ovary inferior; petals absent.*

Leaves gland-dotted, opposite or alternate; stipules absent; stamens numerous; shrubs or trees, rarely herbs *Myrtaceae*
Leaves not gland-dotted or if so the leaves reduced to scales; mostly herbs or herbaceous climbers:
 Green plants with well-developed leaves and not parasitic:
  Leaves alternate or all radical:
   Leaves with stipules, these sometimes adnate to the petiole:
    Flowers bisexual:
     Calyx present; inflorescence not leaf-opposed *Saxifragaceae*
     Calyx absent; upper leaves forming an involucre of bracts; inflorescence leaf-opposed *Saururaceae*
    Flowers unisexual; stamens mostly numerous; ovary and fruit often winged *Begoniaceae*
   Leaves without stipules:
    Ovary often gaping at the top; stamens 4 to many; calyx not unilateral; trees or shrubs; leaves simple or pinnate *Datiscaceae*
    Ovary closed at the top; stamens 6 to many; calyx unilateral or actinomorphic, mostly coloured; climbers or trailers *Aristolochiaceae*
  Leaves opposite, ericoid; flowers crowded *Grubbiaceae*
 Root parasites with scale-like nevergreen leaves growing on the roots of other plants mostly trees and shrubs:
  Flowers bisexual; anthers crowded into 3 or 4 sessile masses; flowers never very large *Hydnoraceae*
  Flowers unisexual; anthers in 1–3 whorls around the column of the gynoecium; flowers sometimes very large *Cytinaceae*

'SYNCARPAE'. **'Axiles'.** Group 12. *Gynoecium composed of* 1 *carpel or of* 2 *or more united carpels with free or united styles, or if carpels free below, then the styles or stigmas united; ovules attached to the central axis or to the base or apex of the ovary; ovary superior, rarely partly immersed in the disk; petals present, free from each other; stamens more or less free from each other and not united into separate bundles, sometimes united into* 1 *bundle (monadelphous), numerous, more than twice the number of the sepals or petals: leaves alternate or all radical.*

Sepals imbricate, contorted, rarely completely connate, or calyptrate (to p. 37):
Anthers narrowly horse-shoe-shaped, the loculi bent on themselves; stipules absent; petals sometimes deeply divided *Gonystylaceae*
Anthers more or less straight:
Petals and stamens perigynous or epigynous:
Leaves with stipules, these sometimes soon falling off or adnate to the petiole:
Seeds with endosperm and a curved embryo:
Herbs mostly with only 2 sepals *Portulacaceae*
Herbs with often very fleshy leaves or shrublets with more than 2 sepals *Aizoaceae (Ficoidaceae)*
Seeds without endosperm or with very little; embryo straight; sepals more than 2:
Ovary with usually more than 2 carpels or sometimes only 1; flowers not or very rarely capitate; style sometimes basal; ovules often 2; filaments sometimes all to one side *Rosaceae*
Ovary of 2 carpels; flowers often capitate; style or styles not basal; stamens not to one side *Hamamelidaceae*
Leaves without stipules:
Stamens free from the petals:
Inflorescence not capitate:
Petals not crumpled in bud; ovules pendulous from the apex of the loculus: *Flacourtiaceae*
Petals often crumpled in bud; ovules basal or from the inner angle of each loculus; sepals valvate *Lythraceae*
Inflorescence capitate; flowers asymmetric; involucre coloured *Hamamelidaceae*
Stamens more or less inserted on the base of the petals *Styracaceae*
Petals and stamens more or less hypogynous or flowers unisexual; disk often present:
Trees, shrubs, or woody climbers (to p. 36):
Leaves compound or rarely unifoliolate and then petiole tumid at the apex (to p. 35):
Leaves pinnate or 3-foliolate (rarely 1-foliolate); petals not calyptrate:
Ovule or ovules ascending or spreading:
Leaves gland-dotted; style or styles central; branches sometimes very spiny *Rutaceae*

Leaves rarely gland-dotted; styles or stigmas often separate:
Wood resinous, bark not bitter; petals as many as sepals *Anacardiaceae*
Wood not resinous; bark not bitter; petals one fewer than the sepals *Sapindaceae*
Wood not resinous; bark very bitter; petals as many as sepals *Simaroubaceae*
Ovule or ovules pendulous:
Stamens free; wood with resin ducts, bark not bitter *Anacardiaceae*
Stamens more or less free; wood without resin ducts; bark very bitter *Simaroubaceae*
Stamens united into a tube often resembling a corolla *Meliaceae*
Leaves digitately 3-5-foliolate; petals calyptrately connate; styles 8–20, filiform; embryo with a large spirally twisted radicle *Caryocaraceae*
Leaves simple, but sometimes deeply lobed (from p. 34):
Leaves with stipules (to p. 36):
Flowers unisexual:
Disk present *Euphorbiaceae*
Disk absent *Flacourtiaceae*
Flowers bisexual:
Torus often enlarged after flowering; ovary mostly deeply lobed, the carpels becoming separated in fruit; anthers sometimes opening by pores *Ochnaceae*
Torus not enlarged and other characters either not present or dis-associated:
Ovary stipitate; sepals more or less connate into a tube *Capparidaceae*
Ovary sessile; sepals usually free:
Sepals 2; stipules split into hairs; stamens numerous *Portulacaceae*
Sepals 3; stipules caducous; flowers cymose or paniculate; petals contorted; stamens 10 or more within a ring of staminodes; ovary 3-locular; Mascarene Isls. *Sarcolaenaceae*
Sepals 4; ovary 2-locular, didymous; disk large, cupular, denticulate; style subgynobasic *Sphaerosepalaceae*
Sepals 5; no staminodes:
Calyx-lobes enlarged and wing-like in fruit:
Flowers mostly rather small and not showy; petals much contorted; calyx-lobes accrescent and wing-like *Dipterocarpaceae*
Flowers showy; ovary 1-locular with a basal placenta *Ochnaceae*
Calyx not enlarged in fruit and not wing-like:
Leaves digitately lobed; flowers large, handsome; anthers opening by short pore-like confluent slits at the apex; petals imbricate or slightly contorted *Cochlospermaceae*
Leaves not lobed:
Stipules large and folded around the terminal bud, intrapetiolar; filaments free *Irvingiaceae*

Stipules not as above, very small or absent:
Petals persistent, contorted; fruit a septicidally dehiscent capsule *Ixonanthaceae*
Petals soon falling off, imbricate-contorted; fruit a drupe *Humiriaceae*
Leaves without stipules (from p. 35):
Ovary and fruit stipitate *Capparidaceae*
Ovary and fruit not stipitate:
Seeds arillate; ovary composed of 1 carpel; leaves mostly with very prominent pinnately parallel lateral nerves; stamens usually persistent *Dilleniaceae*
Seeds not arillate; ovary usually composed of 2 or more carpels:
Sterile flowers with modified pitcher-like saccate or spurred bracts *Marcgraviaceae*
Sterile flowers not present; no modified bracts:
Ovary 1-locular, composed of 1 carpel:
Stamens more than 10, free:
Leaves pellucid-punctate, rather large and coriaceous *Winteraceae*
Leaves not punctate, very small and crowded *Tamaricaceae*
Ovary 2- or more-locular:
Anthers basifixed; seeds usually few, sometimes winged:
Petals contorted:
Stamens connate into a tube ovules usually 2, rarely 1 or more in each loculus *Meliaceae*
Stamens free; ovules usually numerous in each loculus *Bonnetiaceae*
Petals imbricate; anthers sometimes opening by terminal pores *Theaceae*
Anthers dorsifixed, versatile; seeds numerous, small:
Climbers; flowers often unisexual; sepals scarcely imbricate; styles numerous, free; anthers inflexed in bud *Actinidiaceae*
Erect trees and shrubs:
Styles 3–5, free or partly connate; anthers opening by pores *Saurauiaceae*
Style simple, slender, entire or minutely dentate:
Stamens numerous; ovules numerous mostly spreading from central axis *Bonnetiaceae*
Stamens 10 or more; ovules 1–3, pendulous from the top of the loculi *Humiriaceae*
Spiny plants with fasciculate leaves; anthers versatile; seed solitary *Didiereaceae*
Herbs, rarely somewhat woody at the base (from p. 34):
Carpels sunk in the torus; aquatic plants with peltate leaves *Nymphaeaceae*
Carpels not sunk in the torus:
Leaves neither sticky-glandular, ciliate, nor modified into pitchers (to p. 37):
Anthers opening by slits lengthwise (to p. 37):

Stamens quite free among themselves but sometimes slightly adherent to the base of the petals:
Sepals more than 2 (6–8):
Fruits with circumscissile dehiscens at the base *Portulacaceae* (*Lewisia*)
Fruits not circumscissile, opening at the top *Helleboraceae* (*Nigella*)
Fruits capsular their full length *Geraniaceae*
Sepals 2 (rarely 3):
Leaves often densely imbricate, thick and fleshy, with scarious or setose stipules; ovules subaxile *Portulacaceae*
Leaves without stipules; ovules on a basal placenta or on the septa *Papaveraceae*
Stamens connate at the base; leaves pinnate *Oxalidaceae*
Anthers opening by a terminal short pore-like slit (from p. 36) *Cochlospermaceae*
Leaves very sticky-glandular or ciliate with setose teeth, not modified into pitchers *Droseraceae*
Leaves modified into tubes or pitchers; stamens numerous, free; ovary 3-5-locular; ovules numerous *Sarraceniaceae*
Sepals or calyx-lobes valvate or open in bud or very minute (from p. 34):
Anthers 2-locular (to p. 38):
Anthers narrowly horse-shoe-shaped, the loculi bent on themselves *Gonystylaceae*
Anthers more or less straight:
Stamens free or slightly united only at the base (to p. 38):
Petals and stamens hypogynous, or flowers unisexual:
Stipules present:
Flowers bisexual; indumentum often stellate *Tiliaceae* or *Sterculiaceae*
Flowers unisexual; indumentum rarely stellate *Euphorbiaceae*
Stipules absent; calyx cupular or much reduced:
Petals not accrescent in fruit:
Petals valvate; leaves simple; anthers opening by a pore or short slit at the side; petals valvate; seeds with endosperm *Scytopetalaceae*
Petals imbricate; leaves digitate; anthers opening by slits lengthwise; leaves not gland-dotted *Caryocaraceae*
Petals imbricate; leaves gland-dotted *Rutaceae*
Petals accrescent in fruit; leaves simple; anthers opening by a slit *Anacardiaceae* (*Melanorrhoea*)
Petals and stamens perigynous or epigynous:
Leaves with stipules, these mostly paired; ovary 2-locular; styles subulate, free *Hamamelidaceae*
Leaves without stipules:
Anthers inflexed in bud; calyx tubular; petals often clawed and crumpled in bud *Lythraceae*
Anthers erect in bud:
Leaves simple; wood not resinous (to p. 38):

Ovary incompletely septate, wholly superior; ovules pendulous from a free central placenta *Olacaceae*

Ovary completely septate, partly inferior *Styracaceae*

Leaves compound or unifoliolate; wood resinous (from p. 37) *Anacardiaceae*

Stamens more or less united into a tube or androphore or into bundles, hypogynous; indumentum usually stellate, more rarely lepidote:

Anthers opening their full length by slits *Sterculiaceae*

Anthers opening by short pore-like slits *Scytopetalaceae*

Anthers 1-locular; stamens monadelphous or in bundles; calyx with or without an epicalyx of bracteoles; indumentum often stellate or more rarely lepidote (from p. 37):

Trees or rarely shrubs; leaves digitately compound or simple; carpels in fruit not or rarely splitting away from the central axis; pollen mostly smooth; styles as many as carpels, or style 1 *Bombacaceae*

Mostly herbs; leaves simple; carpels often splitting away from the central axis or becoming free in fruit (schizocarpic); styles as many or twice as many as the carpels; pollen more or less muricate *Malvaceae*

'SYNCARPAE'. **'Axiles'.** Group 13. *Gynoecium composed of* 1 *carpel or of* 2 *or more united carpels with free or united styles, or if carpels free below, then the styles or stigmas united; ovules attached to the central axis or to the base or apex of the ovary; ovary superior, rarely partly immersed in the disk; petals present, free from each other; stamens more or less free from each other and not united into several separate bundles, sometimes united into* 1 *bundle (monadelphous), numerous, more than twice the number of the sepals or petals; leaves opposite or verticillate.*

Calyx-lobes or sepals imbricate or calyptrate (to p. 40):
Sepals mostly 2, caducous:
Petals 4, often crumpled in bud; sepals imbricate; filaments free; ovules more than 1 — *Papaveraceae*
Petals 4–6, not crumpled in bud; filaments free or connate at the base; ovule 1 in each loculus; sepals valvate — *Clusiaceae* (*Guttiferae*)
Sepals more than 2, mostly persistent; petals usually 5:
Stipules absent:
Leaves digitately 3-5-foliolate; flowers in terminal racemes; bracts none — *Caryocaraceae*
Leaves simple:
Stamens if united into bundles then these connate nearly to the top, rarely simply monadelphous; style single — *Hypericaceae*
Stamens free or the anthers united into a mass; flowers sometimes dioecious or polygamous:
Petals clawed and crumpled in bud — *Lythraceae*
Petals neither clawed nor crumpled in bud:
Herbs or shrublets with often thick fleshy leaves, not gland-dotted — *Aizoaceae* (*Ficoidaceae*)
Spiny fleshy shrubs; leaves fasciculate — *Didiereaceae*
Unarmed trees and shrubs:
Leaves glandular-punctate — *Rutaceae*
Leaves not punctate:
Styles several in a ring on the shoulders of the carpels; capsule septicidally dehiscent from the base and the valves diverging like the ribs of an umbrella; flowers bisexual — *Medusagynaceae*
Styles not as above, often simple and little branched; fruit dehiscent or not; flowers mostly unisexual, polygamous or dioecious — *Clusiaceae* (*Guttiferae*)
Stipules present; sometimes these spinescent, rarely divided into hairs:
Stipules split into hairs; sepals 3 — *Portulacaceae*
Stipules not split into hairs:
Leaves simple: (see also *Eucryphiaceae* p. 40):
Leaves with 3 or more longitudinal parallel main nerves parallel with the margin — *Melastomataceae*

Leaves pinnately nerved:
Stamens 15 or more, free or very shortly connate at the base; styles 2 or 3, free; leaves with numerous very fine transverse (feather veined) tertiary nerves; stipules mostly paired, lateral, rigid; flowers paniculate or racemose *Quiinaceae*
Stamens very numerous, free; styles 5–12; ovary with as many loculi; leaves simple or pinnate with small caducous intrapetiolar stipules; flowers solitary, showy; seeds winged *Eucryphiaceae*
Stamens numerous; style 1; stigma 3-lobed *Theaceae*
Stamens 10–15, free; style with 2–6 stigmas; leaves rather fleshy, with small persistent stipules; flowers in cymes *Zygophyllaceae*
Leaves digitately 3-foliolate; flowers in terminal ebracteate racemes; stamens very numerous; styles filiform *Caryocaraceae*
Calyx-lobes or sepals valvate (from p. 39):
Stamens free or very shortly connate at the base:
Stamens neither inflexed nor reflexed in bud:
Stipules in pairs, not between the petioles, minute; stamens hypogynous *Tiliaceae*
Stipules single, between the petioles; stamens perigynous or epigynous *Rhizophoraceae*
Stipules absent; stamens perigynous; leaves simple *Philadelphaceae*
Stipules absent; stamens hypogynous; leaves trifoliolate *Oxystylidaceae**
Stamens inflexed or reflexed in bud:
Leaves stipulate; indumentum often stellate or lepidote; anthers usually basifixed *Euphorbiaceae*
Leaves without stipules; indumentum rarely stellate; anthers versatile:
Flowers sometimes large and showy; petals clawed, often crumpled in bud; ovary few-locular; branches sometimes spinescent *Lythraceae*
Flowers with small subsessile petals; ovary many to few-locular; stamens numeruos; trees *Sonneratiaceae*
Stamens monadelphous or in fascicles opposite the petals, the latter contorted or imbricate; leaves usually stipulate; indumentum often stellate or lepidote *Sterculiaceae*

'SYNCARPAE'. **'Axiles'.** Group 14. *Gynoecium composed of* 1 *carpel or of* 2 *or more united carpels with free or united styles, or if carpels free below, then the styles or stigmas united; ovules attached to the central axis or to the base or apex of the ovary; ovary superior, rarely partly immersed in the disk; petals present, free from each other; stamens more or less free from each other and not united into separate bundles, sometimes united into* 1 *bundle (monadelphous); flowers actinomorphic ('regular'); perfect stamens the same number as the petals and alternate with them, or up to twice as many; leaves simple (not unifoliolate), alternate or all radical.*

Leafless parasitic nevergreen herbs on roots of trees *Monotropaceae*
Leafy trees or shrubs, but sometimes leafless at time of flowering:
 Anthers opening by valves:
  Ovary composed of 2 carpels; stipules often present, mostly paired; flowers often capitate; petals often linear and circinate in bud *Hamamelidaceae*
  Ovary composed of 1 carpel; stipules absent *Lauraceae*
 Anthers opening by apical pores or pore-like slits (not to the base):
  Leaves with 3 or more longitudinally parallel main nerves; connective of the anthers often produced at the base *Melastomataceae*
  Leaves pinnately nerved or nerves obscure:
   Petals imbricate or contorted:
    Ovary deeply vertically lobed; style often basal between the lobes:
     Torus enlarging in fruit and the carpels often becoming separate; ovules 1 or 2 in each loculus *Ochnaceae*
     Torus not enlarged; herbs; leaves mostly radical and evergreen, rounded *Pyrolaceae*
     Torus not enlarged; shrublets *Euphorbiaceae*
    Ovary not deeply lobed and torus not enlarged; style more or less terminal:
     Ovules numerous to several in each loculus or in the ovary:
      Anthers inflexed in bud; stamens 10-12 *Clethraceae*
      Anthers not inflexed in bud; stamens 5 *Pittosporaceae*
     Ovules 1-3 in each loculus or in the ovary:
      Branches, leaves and inflorescence lepidote all over *Aextoxicaceae*
      Branches etc. not lepidote:
       Fruits winged: leaves not glandular *Cyrillaceae*
       Fruits not winged:
        Leaves very glandular-pilose, circinate in bud *Byblidaceae*
        Leaves not glandular and not circinate in bud *Pentaphylacaceae*
   Petals induplicate-valvate; slender heath-like shrublets; flowers axillary, solitary *Tremandraceae*
 Anthers opening by slits lengthwise to the base:
  Trees, shrubs, shrublets or woody climbers (to p. 46):
   Leaves with stipules (stipules often soon falling off and leaving a scar) or very minute (to p. 43):

Calyx-lobes persistent, accrescent and wing-like in fruit; petals contorted; cotyledons folded around the radicle:
Erect trees and shrubs; ovary 3-locular, ovules 2 in each locolus *Dipterocarpaceae*
Climbing shrubs with hooked branches; ovary 1-locular with 1 ascending ovule *Ancistrocladaceae*
Calyx not wing like in fruit; petals rarely contorted:
Flowers bisexual (to p. 43):
Stamens perigynous, inserted on the calyx-tube:
Style near the base of the ovary or on its side; ovary 1- or 2-locular, composed of 1 or 2 carpels *Rosaceae*
Style or styles not basal or lateral:
Ovary 2-locular *Hamamelidaceae*
Ovary imperfectly 3-locular at the base, 1-locular at the apex; ovules borne at the apex of a slender free basal placenta; flowers in a dense globose umbel *Dipentodontaceae*
Stamens hypogynous or inserted on or at the base of a disk:
Stipules axillary (intrapetiolar), folded in bud, often very long but usually soon caducous and leaving a scar:
Trees and shrubs:
Petals not appendaged inside:
Flowers solitary; sepals 10-8 *Strasburgeriaceae*
Flowers in cymes, racemes or panicles:
Petals contorted, persistent; stamens 2-4-times as many as the petals *Ixonanthaceae*
Petals imbricate; stamens 10 *Irvingiaceae*
Petals appendaged inside, imbricate, at length deciduous; stamens twice as many as petals *Erythroxylaceae*
Herbs with terminal spike-like racemes *Linaceae*
Stipules not axillary:
Disk present, annular or of separate glands:
Petals entire or emarginate; ovules erect or ascending:
Petals imbricate *Celastraceae*
Petals valvate, linear from a broader base, upper parts inflexed in bud; anthers subsessile, with setose-pilose connective *Goupiaceae*
Petals often deeply lobed; ovules pendulous *Dichapetalaceae* (*Chailletiaceae*)
Disk absent (the torus sometimes enlarging in fruit but not disk-like):
Sepals 3, very imbricate; petals imbricate-contorted; stamens inserted inside a cup of staminodes; Mascarene Isls. *Sarcolaenaceae*
Sepals 5 or 4; no cup of staminodes present:
Sepals imbricate; anthers 2-locular:
Stamens 10, in 1 series:
Hairs of the leaves often medifixed; petiole often glandular like the sepals; petals often clawed *Malpighiaceae*
Hairs not medifixed; petiole and sepals not glandular; fertile stamens often 5 with 5 infertile *Linaceae*

Stamens 10 in 2 series; hairs when present not medifixed; petiole and calyx not glandular *Erythroxylaceae*
Stamens usually numerous; hairs not medifixed *Ochnaceae*
Stamens 4 or 5 *Aquifoliaceae*
Sepals valvate:
Anthers 4-locular; hairs when present not medifixed *Huaceae*
Anthers 2-locular; hairs often medifixed *Malpighiaceae*
Flowers unisexual, sometimes polygamo-dioecious (from p. 42):
Disk absent; petals not bilobed:
Stamens hypogynous; seeds with endosperm; style never basal
Ovules more than 1 in each ovary-loculus, axile *Flacourtiaceae*
Ovule 1 in each ovary-loculus, pendulous *Pandaceae*
Stamens perigynous or epigynous; seeds without endosperm; style sometimes basal *Rosaceae*
Disk present, sometimes of hypogynous glands:
Petals bilobed *Dichapetalaceae*
Petals not bilobed:
Stipules conspicuous, persistent *Euphorbiaceae*
Stipules very inconspicuous, caducous *Celastraceae*
Leaves without stipules (from p. 41):
Stamens united into a hollow tube often resembling a corolla:
Leaves not gland-dotted *Meliaceae*
Leaves gland-dotted *Rutaceae*
Stamens united into a central column *Aptandraceae*
Stamens free or connate only at the base:
Stamens hypogynous or very slightly perigynous (rarely loosely adnate to the petals) (to p. 45):
Sterile flowers present with modified pitcher-like bracts; mostly epiphytes *Marcgraviaceae*
Sterile flowers not as above or absent:
Stamens double the number of the petals or fewer only by the abortion of some anthers (to p. 44):
Sepals usually with 2 large glands outside at the base; indumentum usually of medifixed hairs *Malpighiaceae*
Sepals not glandular at the base; hairs rarely medifixed:
Ovary 1-locular (sometimes so only at the top) (to p. 44):
Ovary of 1 carpel:
Torus forming a stipe:
Ovule pendulous from the apex of the loculus or from a basal funicle; pedicels not adnate to the bract *Anacardiaceae*
Ovule erect from the base of the loculus; pedicels adnate to the large veiny bract *Podoaceae*
Torus not forming a stipe *Phytolaccaceae*
Ovary of more than 1 carpel:
Leaves very small; flowers spicate-racemose *Tamaricaceae*
Leaves not very small:
Ovary 1-locular with 1 free central placenta and basal ovules *Stegnospermataceae*

Ovary with ovules pendulous from the top of a free central placenta *Olacaceae*

Ovary completely 2- or more-locular (from p. 43):

Disk absent:

Leaves reduced or scale-like; placenta basal *Molluginaceae*

Leaves not as above; placentas axile or apical:

Flowers unisexual; ovary 3- or 4-locular:

Petals of the male flowers imbricate *Pentadiplandraceae*

Petals of the male flowers valvate *Pandaceae*

Flowers bisexual:

Branches stiff and spine-tipped; leaves minute and scale-like; fruit a capsule or berry; embryo straight or much curved *Koeberliniaceae*

Branches not spine-tipped; leaves normally developed; fruit a berry; embryo straight *Stachyuraceae*

Disk present:

Bark not bitter *Olacaceae*

Bark very bitter *Simaroubaceae*

Stamens the same number as the petals or fewer (from p. 43).

Disk absent;

Peduncle adnate to the petiole; carpels elevated on a gynophore; embryo hook-like *Cneoraceae*

Peduncle free from the petiole; carpels not elevated on a gynophore:

Ovules numerous; anthers opening lengthwise or by apical pores: style simple; embryo minute in copious endosperm *Pittosporaceae*

Ovules 6 in each of the 2 loculi of the ovary *Koeberliniaceae*

Ovules 1 or 2 in each loculus, pendulous:

Anthers opening by an apical pore *Theaceae*

Anthers opening by slits lengthwise:

Petals clawed, imbricate or contorted *Linaceae*

Petals not clawed, valvate or slightly imbricate:

Ovary 3- or more-locular; seeds with straight embryo *Aquifoliaceae*

Ovary 1- or 2-locular:

Ovules pendulous from the top of the ovary; seeds with more or less straight embryo *Icacinaceae*

Ovule 1, pendulous from the central axis; seeds with a horse-shoe-shaped embryo *Pellicieraceae*

Ovule 1 in each loculus, basal *Tetrameristaceae*

Disk present, sometimes of separate glands, or inconspicuous:

Petals valvate; ovules pendulous:

Stamens opposite the petals; flowers mostly bisexual *Olacaceae*

Stamens alternate with the petals; disk sometimes unilateral; flowers often dioecious *Icacinaceae*

Petals imbricate or contorted:

Leaves gland-dotted *Rutaceae*
Leaves not gland-dotted but sometimes lepidote:
Petals more or less spreading, not connivent:
Stamens usually 5:
Leaves densely lepidote *Aextoxicaceae*
Leaves not lepidote:
Ovules 1-3 in each loculus:
Ovule 1; wood resinous *Anacardiaceae*
Ovules mostly 2 in each loculus: wood not resinous *Celastraceae*
Ovules several to many in each loculus *Pittosporaceae*
Stamens 3, with flattened filaments; wood not resinous; flowers cymose *Hippocrateaceae*
Petals erect and more or less connivent:
Petals connivent only in the upper part; small herbs with a woody branched rhizome; anthers opening by 2 slits *Stackhousiaceae*
Petals free in the upper part; shrubs or small trees; anthers opening by a single slit *Epacridaceae*
Stamens very distinctly perigynous or epigynous (from p. 43):
Leaves with 3 or more longitudinally parallel nerves *Melastomataceae*
Leaves with more or less pinnate or palmate nervation:
Carpels free at the apex; leaves stipulate *Hamamelidaceae*
Carpels completely united; stipules absent:
Anthers inflexed in bud; calyx-lobes often with accessory lobes between *Lythraceae*
Anthers erect in bud:
Petals scale-like, opposite the sepals; ovary wholly superior:
Fruit a capsule *Aquilariaceae*
Fruit indehiscent:
Calyx tubular, sometimes petaloid; corolla of small scales *Thymelaeaceae*
Calyx small and 4- or 5- toothed; corolla conspicuous *Icacinaceae*
Petals not scale-like, alternate with the sepals:
Ovules axile:
Ovary wholly superior; petals more or less clawed; stamens only slightly perigynous; plants woody only at the base *Molluginaceae*
Ovary semi-inferior (or very rarely superior); stamens very clearly perigynous; flowers often in heads; low shrubs with small leaves *Bruniaceae*
Ovary quite or semi-superior; trees or shrubs with fairly large leaves sometimes with spinulose or glandular-serrate margins *Escalloniaceae*
Ovules inserted at the top of a central placenta free at the top *Olacaceae*

Herbs, rarely slightly woody at the base (from p. 41):
Nevergreen parasites destitute of chlorophyll; leaves reduced to scales *Monotropaceae*
Not parasitic; leaves green:
Leaves densely covered with sticky gland-tipped processes or setose-ciliate and bilobed, stipulate, often circinate in bud *Droseraceae*
Leaves not so glandular and other characters not associated:
Leaves with stipules and these never gland-like:
Stamens and petals hypogynous, sometimes flowers unisexual:
Flowers bisexual:
Sepals more than 2:
Sepals valvate *Tiliaceae*
Sepals imbricate:
Ovary more or less deeply lobed; calyx if spurred then spur concealed and adnate to the pedicel; leaves rarely unequal (anisophyllous) *Geraniaceae*
Ovary entire; calyx never spurred:
Petals contorted; stipules not scarious *Linaceae*
Petals imbricate or small; stipules scarious *Molluginaceae*
Sepals 2, very much imbricate *Portulacaceae*
Flowers unisexual *Euphorbiaceae*
Stamens and petals perigynous; anthers inflexed in bud *Lythraceae*
Leaves without stipules or these represented by glands:
Sepals 2, much imbricate *Portulacaceae*
Sepals more than 2:
Petals connivent in the upper part; small herbs from a woody branched rhizome; ovule basal, solitary, erect in each loculus *Stackhousiaceae*
Petals not connivent:
Petals and stamens hypogynous or very slightly perigynous:
Style gynobasic, the carpels free or nearly so; leaves much dissected; ovules ascending *Limnanthaceae*
Style not gynobasic:
Stigmas sessile and alternating with the carpels (commissural) *Francoaceae*
Styles or stigmas continuous with the carpels (loculi of the ovary):
Petals contorted, often fugacious, mostly large and conspicuous; leaves entire or nearly so *Linaceae*
Petals imbricate, often very small and inconspicuous; fruits sometimes winged; leaves not gland-dotted *Molluginaceae*
Petals imbricate; leaves gland-dotted *Rutaceae*
Petals and stamens perigynous:
Anthers 1-locular, peltate; leaves radical and long-petiolate, compound; flowers in a head *Adoxaceae*
Anthers 2-locular:
Anthers inflexed in bud; seeds without endosperm *Lythraceae*
Anthers erect in bud:
Calyx-tube adnate to the lowermost third of the ovary; ovule 1

in each loculus at the base of the central axis; S.W. Australia *Eremosynaceae*

Calyx-tube free from the ovary; ovules several to numerous in each loculus or carpel; wide distribution:

Seeds with endosperm; carpels without a gland or scale at the base *Saxifragaceae*

Seeds without endosperm; carpels with a gland or scale at the base *Crassulaceae*

'SYNCARPAE'. **'Axiles'.** Group 15. *Gynoecium composed of* 1 *carpel or of* 2 *or more united carpels with free or united styles; or if carpels free below, then the styles or stigmas united; ovules attached to the central axis or to the base or apex of the ovary; ovary superior, rarely partly immersed in the disk; petals present, free from each other; stamens more or less free from each other and not united into separate bundles, sometimes united into* 1 *bundle (monadelphous); flowers actinomorphic ('regular'); perfect stamens the same number as the petals and alternate with them or not more than twice as many, rarely fewer; leaves simple, opposite or verticillate or rarely fasciculate, never all radical or completely reduced.*

Ovary stipitate on a distinct gynophore:
 Leaves not gland-dotted:
  Ovules inserted on the intrusive septa and appearing to be axile *Capparidaceae*
  Ovules inserted on the central axis or at the base of the ovary *Caryophyllaceae*
 Leaves gland-dotted *Rutaceae*
Ovary sessile or rarely very shortly stipitate:
 Trees, shrubs, shrublets, or woody climbers (to bottom of p. 50):
  Leaves with stipules, these sometimes rudimentary or consisting of hairs rarely by a narrow rim: (to p. 49):
   Stipules intrapetiolar, often connivent into one or adnate to the petiole:
    Sepals not glandular outside and not accrescent in fruit; hairs when present not medifixed *Erythroxylaceae*
    Sepals usually with a pair of conspicuous glands outside, sometimes accrescent in fruit; hairs medifixed; leaves often biglandular at the base or on the petiole *Malpighiaceae*
   Stipules not intrapetiolar, sometimes rudimentary:
    Disk absent or inconspicuous or of separate glands; calyx often glandular:
     Stamens free or shortly united only at the base (to top of p. 49):
      Calyx mostly with a pair of glands near the base; trees, shrubs or climbers; petals often clawed; filaments without a scale at the base *Malpighiaceae*
      Calyx not glandular; anther-loculi collateral:
       Flowers bisexual:
        Filaments of the stamens often with a scale at the base; stamens hypogynous *Zygophyllaceae*
        Filaments without a scale at the base:
         Stamens perigynous *Cunoniaceae*
         Stamens hypogynous *Vivianiaceae*
       Flowers unisexual; indumentum stellate *Hamamelidaceae*
      Calyx not glandular; anther-loculi back to back; trees or shrubs; filaments without a scale; petals 4; stamens 4 *Salvadoraceae*

Stamens united into a long tube; sepals not glandular; ovary 5-locular with 2 ovules in each; calyx valvate *Rhizophoraceae*
Disk present and conspicuous; calyx not glandular:
Flowers bisexual:
Stamens inserted on or below the margin of the disk; filaments subulate:
Stamens 3-5 *Celastraceae*
Stamens 8-10:
Sepals imbricate; filaments inserted on the inside of the disk; seed pendulous from a long funicle and with a fibrous aril *Ctenolophonaceae*
Sepals valvate or rarely imbricate; filaments on the edge or on the outside of the disk:
Style mostly simple; stamens often in pairs opposite the petals; ovary mostly inferior *Rhizophoraceae*
Styles free; stamens not in pairs; ovary superior *Cunoniaceae*
Stamens inserted on the surface of the disk; filaments flattened or connivent, often adnate to the ovary; petals imbricate or valvate *Hippocrateaceae*
Flowers unisexual; ovules pendulous from the apex of each loculus:
Seeds often carunculate; flowers sometimes but rarely in heads *Euphorbiaceae*
Seeds not carunculate; flowers arranged in dense heads; hairs stellate *Hamamelidaceae*
Leaves without stipules or these represented by glands (from p. 48):
Stamens united into a tube:
Stamens more than 4; flowers bisexual *Meliaceae*
Stamens 4; flowers unisexual *Clusiaceae*
Stamens free or very shortly united at the base:
Anthers opening all over by several minute pores *Icacinaceae*
Anthers opening only at the apex by one or two pores; stamens as many or twice as many as the petals:
Leaves often with 3-9 longitudinally parallel main nerves; anthers often appendaged at the base *Melastomataceae*
Leaves with a single marginal nerve; anthers not appendaged at the base; slender heath-like shrublets, sometimes with winged stems and reduced foliage *Tremandraceae*
Anthers opening lengthwise by slits:
Ovules numerous in each loculus of the ovary (to p. 50):
Flowers bisexual (to top of p. 50):
Petals and stamens hypogynous:
Stamens 10; embryo large, the endosperm very scanty *Ledocarpaceae*
Stamens 5; embryo small, with copious endosperm *Pittosporaceae*
Petals and stamens perigynous:
Calyx often with teeth between the lobes; petals often crumpled in bud *Lythraceae*
Calyx without teeth between the lobes; petals not crumpled in bud *Philadelphaceae*

Flowers unisexual; petals and stamens hypogynous; stamens mostly numerous *Clusiaceae*
Ovules few or 1 in each loculus of the ovary (from p. 49):
Sepals 2-glandular outside; hairs on the leaves (when present) often medifixed:
Fruits usually winged (samaroid) or covered with long bristles; styles often 2 *Malpighiaceae*
Fruits drupaceous, not winged; style 1 *Simaroubaceae*
Sepals not glandular outside; hairs (when present) not medifixed:
Ovules pendulous from the apex of the loculi of the ovary, or axile:
Sepals imbricate or open in bud:
Sepals free or calyx-tube very short:
Stamens hypogynous:
Petals imbricate *Linaceae*
Petals valvate *Icacinaceae* and a few *Oleaceae*
Stamens perigynous:
Ovary 5-locular; ovules 2 in each loculus *Escalloniaceae*
Ovary 4-locular; ovule solitary *Cunoniaceae*
Sepals connate into a long or rather long tube *Thymelaeaceae*
Sepals valvate:
Stamens hypogynous; seeds with endosperm:
Petals contorted; embryo much curved or circinate *Vivianiaceae*
Petals induplicate-valvate; embryo straight, small or minute *Tremandraceae*
Stamens perigynous; fruits mostly winged; seeds without endosperm *Combretaceae*
Ovules erect or ascending from or near the base of the loculi:
Calyx imbricate:
Filaments subulate or filiform, not flattened:
Leaves mostly palmately lobed, or if not so then disk absent; ovary and fruit compressed at a right angle to the septum *Aceraceae*
Leaves pinnately lobed; disk present:
Flowers bisexual *Celastraceae*
Flowers dioecious or polygamous *Clusiaceae*
Filaments flattened *Hippocrateaceae*
(See also *Turpinia* (*Staphyleaceae*))
Calyx valvate; stamens 3-5; style short or stigmas sessile:
Fruits united and forming a fleshy ovoid mass *Batidaceae*
Fruits not united:
Leaves not gland-dotted:
Sepals free or nearly so *Anacardiaceae* (*Bouea*)
Sepals united into a long tube, the 2 posticuos petals often larger than the others *Lythraceae*
Sepals united into a short tube; petals equal *Podoaceae*
Leaves gland-dotted *Rutaceae*
Herbs, sometimes rather woody only at the base (from p. 48):
Ovary incompletely septate, with free central or basal placentation (to p. 51):

Sepals or calyx-lobes the same number as the petals, 3-5; petals 3-5, often clawed, hypogynous; leaves connected at the base by a transverse line or stipulate *Caryophyllaceae*

Sepals the same number as the petals but often with as many alternate accessory lobes; stamens more or less perigynous, anthers inflexed in bud; petals often crumpled in bud; some aquatics *Lythraceae*

Sepals fewer than the petals, usually 2; petals 4 or 5 or more; stamens opposite the petals *Portulacaceae*

Ovary completely septate or nearly so into separate loculi (from p. 50:)

Leaves with stipules, these not gland-like:

Stipules in pairs:

Branches not jointed at the nodes; ovules numerous, attached to the central axis; leaves sometimes verticillate *Elatinaceae*

Branches usually not jointed; ovules few, often only 2, pendulous; sepals often unequal; petals usually with alternate glands; leaves sometimes unequal (anisophyllous) *Geraniaceae*

Branches jointed; ovules few, pendulous or ascending; sepals equal *Zygophyllaceae*

Stipules single, sometimes minute:

Stamens distinctly perigynous; capsule circumscissile *Aizoaceae* (*Ficoidaceae*)

Stamens hypogynous or only slightly perigynous:

Ovules pendulous; fruit a capsule or drupe, not circumscissile *Linaceae*

Ovules basal or axile and spreading; fruit a valvular capsule, not circumscissile *Molluginaceae*

Leaves without stipules or rarely these represented by glands or by a transverse line:

Leaves with 3 or more longitudinally parallel nerves; anthers usually appendaged and opening by a terminal pore *Melastomataceae*

Leaves without nerves as above; anthers opening lengthwise by slits:

Stamens hypogynous or nearly so:

Petals not contorted, very small; seeds with curved embryo; fruit loculicidally dehiscent or by a transverse slit *Molluginaceae*

Petals contorted:

Seeds with straight embryo; fruit septicidally dehiscent *Linaceae*

Seeds with much curved or circinate embryo; fruit loculicidally dehiscent *Vivianiaceae*

Stamens perigynous; anthers erect in bud *Aizoaceae* (*Ficoidaceae*)

Stamens perigynous; anthers inflexed in bud:

Ovary superior; terrestrial or semi-aquatic herbs; cotyledons equal *Lythraceae*

Ovary semi-inferior; floating aquatic herbs with leaves of two kinds, the lower much divided, fruit spiny *Trapaceae*

'SYNCARPAE'. **'Axiles'.** Group 16. *Gynoecium composed of* 1 *carpel or of* 2 *or more united carpels with free or united styles or if carpels free below, then the styles or stigmas united; ovules attached to the central axis or to the base or apex of the ovary; ovary superior, rarely partly immersed in the disk; petals present, free from each other; stamens more or less free from each other and not united into separate bundles, sometimes united into* 1 *bundle (monadelphous); flowers actinomorphic; perfect stamens the same number as the petals and alternate with them or twice as many or fewer; leaves compound, rarely unifoliolate and then petiole distinctly tumid at the apex.*

Leaves alternate or all radical (to p. 53):
- **Stamens** united into a tube; leaves pinnate or rarely unifoliolate:
  - **Leaves** without stipules — *Meliaceae*
  - **Leaves** with intrapetiolar stipules — *Melianthaceae*
- **Stamens** free or united only at the base:
  - **Leaves** with stipules:
    - **Stipules** intrapetiolar, sometimes very long and wrapped around the terminal bud, often soon falling off:
      - **Ovules** pendulous; flowers actinomorphic ('regular') — *Irvingiaceae*
      - **Ovules** ascending; flowers more or less zygomorphic ('irregular') — *Melianthaceae*
    - **Stipules** not intrapetiolar:
      - **Trees,** shrubs, or climbers; leaves sometimes unifoliolate:
        - **Leaves** simple or very rarely pinnate; seeds not arillate; ovary excentric — *Ochnaceae*
        - **Leaves** 3-foliolate or 1-foliolate; seeds arillate; disk absent; ovary central — *Lepidobotryaceae*
        - **Leaves** simple or compound; seeds often arillate; ovary central — *Sapindaceae*
      - **Herbs,** or slightly woody; leaves digitately or pinnately compound:
        - **Stamens** with a scale at the base of the filaments — *Zygophyllaceae*
        - **Stamens** without a scale at the base of the filaments:
          - **Ovules** 1 or 2 in each loculus, mostly pendulous;
            - **Stipules** not fimbriate; ovary not stipitate — *Geraniaceae*
            - **Stipules** fimbriate; ovary stipitate — *Oxystylidaceae**
          - **Ovules** numerous in each loculus; ovary not stipitate — *Saxifragaceae*
  - **Leaves** without stipules:
    - **Leaves** gland-dotted; disk usually present and within the stamens — *Rutaceae*
    - **Leaves** not gland-dotted:
      - **Leaves** 2-foliolate; shrubs or small trees with axillary simple or forked spines; ovary semi-immersed in the disk; fruit a fleshly oily drupe — *Balanitaceae*
      - **Leaves** not 2-foliolate:
        - **Ovules** pendulous from towards the apex of each loculus (to p. 53):

Ovules 6-12 in each loculus, 2-seriate; flowers small, paniculate; large trees with coloured wood; leaves imparipinnate *Meliaceae*
Ovule solitary or 2 and collateral:
Ovary of more than 1 carpel:
Ovary entire; stamens without a scale at the base; ovules usually 2 in each loculus *Burseraceae*
Ovary mostly deeply lobed or carpels free; stamens often with a scale at the base; ovule usually solitary in each loculus *Simaroubaceae*
Ovary of 1 carpel; wood resinous; ovule solitary *Anacardiaceae*
Ovules in pairs and superposed *Akaniaceae*
Ovules ascending or horizontal and axile:
Ovules numerous, spreading horizontally:
Leaves trifoliolate; sepals, petals, and stamens 8, rarely 5; stigmas sessile, 8-rayed *Tovariaceae*
Leaves imparipinnate; sepals 5; petals 5; stamens 10 or 5; fruit a large oblong berry *Averrhoaceae*
Ovules mostly few or 1:
Herbs often with sensitive digitate or pinnate leaves:
Styles 5, free, persistent; seeds with copious endosperm *Oxalidaceae*
Style 1; seeds without endosperm:
Ovary sessile *Limnanthaceae*
Ovary long-stipitate *Oxystylidaceae**
Trees or shrubs:
Ovules about 8 in each loculus; style 1 *Meliaceae*
Ovules 2 in each carpel or loculus, collateral *Connaraceae*
Ovule 1 in each loculus, erect; leaves pinnate or unifoliolate *Sapindaceae*

Leaves opposite (rarely subopposite) or verticillate, or rarely fasciculate, never all radical or completely reduced (from p. 52):
Stamens numerous, more than twice the number of the petals:
Leaves digitate:
Leaves large; flowers in terminal racemes *Caryocaraceae*
Leaves very small, sessile, 3-foliolate; flowers axillary, solitary; anthers didymous, opening by pore-like slits; ovary 2-locular *Baueraceae*
Leaves pinnate; petals and stamens subhypogynous *Eucryphiaceae*
Stamens definite in number, not more than twice as many as the petals; disk usually present:
Leaves gland-dotted; disk usually present between the stamens and ovary; ovary often deeply lobed; loculi 2-ovuled; ovules pendulous; stipules rarely present *Rutaceae*
Leaves not gland-dotted (rarely lepidote), sometimes fleshy; sometimes glandular-serrate:
Leaves stipulate, stipules often deciduous (to p. 54):
Ovules pendulous from the central axis, 2 or more in each loculus; disk often fleshy, rarely absent:
Stipules persistent and often paired; shrubs or herbs; leaves mostly

2-foliolate or pinnate; filaments often with a scale or gland attached to the base *Zygophyllaceae*

Stipules often deciduous, sometimes large; leaves 3- or more-foliolate, rarely lepidote; disk perigynous; filaments without a scale at the base; leaves sometimes glandular-serrate *Cunoniaceae*

Ovule pendulous, solitary in each loculus; disk usually within the stamens; trees or shrubs; leaves mostly pinnate; filaments nude, pilose, or with a scale at the base *Simaroubaceae*

Ovules ascending; stamens inserted outside the disk; leaves pinnate; fruits sometimes inflated *Staphyleaceae*

Leaves without stipules (from p. 53):

Filaments of the stamens free or nearly so:

Leaves pinnate:

Ovary and fruit compressed at right angles to the septum; ovules 2 in each loculus, collateral *Aceraceae*

Ovary and fruit not compressed at right angles to the septum:

Stamens fewer than the petals; style simple; anther-loculi back to back; ovules usually 2 in each loculus *Oleaceae*

Stamens as many as or more than the petals:

Ovule 1 in each loculus:

Petals one less than the sepals; bark not bitter *Sapindaceae*

Petals as many as the sepals; bark very bitter *Simaroubaceae*

Ovules mostly 2 in each loculus; stamens more than 5 *Burseraceae*

Ovules 8 or more in each loculus:

Stamens more than 5 *Meliaceae*

Stamens 5 *Staphyleaceae*

Leaves digitately 3-foliolate, sessile; small shrubs with very small seeds *Baueraceae*

Leaves digitately 5-9-foliolate, petiolate; large trees; sepals free or united into a tube; seeds large, with a broad hilum; petals unequal, clawed *Hippocastanaceae*

Filaments of the stamens more or less connate into a tube, often double the number of the petals *Meliaceae*

'SYNCARPAE'. **'Axiles'.** Group 17. *Gynoecium composed of* 1 *carpel or of* 2 *or more united carpels with free or united styles, or if carpels free below, then the styles or stigmas united; ovules attached to the central axis or to the base or apex of the ovary; ovary superior, rarely partly immersed in the disk; petals present, free from each other; perfect stamens the same number as the petals and opposite to them (very rarely fewer).*

Leaves not gland-dotted; petals and stamens more or less hypogynous or subperigynous; disk usually conspicuous, sometimes much enlarged in fruit (to p. 56):
Calyx-lobes or free sepals imbricate, well developed:
Petals imbricate or biseriate:
Ovary 1-locular; ovules basal or on a basal placenta:
Petals often biglandular towards the base; ovules few; leaves simple or compound; anthers opening by valves; shrubs often with long and short shoots, usually spinous *Berberidaceae*
Petals not glandular at the base; anthers opening by longitudinal slits; herbs or fleshy shrubs:
Ovules more than 1 in the ovary; sepals more than 2, usually 5; herbs; stipules absent *Primulaceae*
Ovules sometimes 1 in the ovary, usually more; sepals usually 2; fleshy shrubs or herbs; stipules mostly present and often scarious; stamen rarely only 1; stem-leaves sometimes opposite *Portulacaceae*
Ovary 2- or 3-locular; ovule solitary, pendulous from near the top or if 2 then ovules collateral:
Sepals free or only slightly united:
Flowers bisexual:
Stamens alternating with petaloid staminodes *Corynocarpaceae*
No staminodes; shrubs or small trees:
Petals alternate with the sepals *Flacourtiaceae*
Petals and stamens opposite the sepals *Sabiaceae*
Flowers dioecious; leaves compound *Simaroubaceae*
Sepals united high up; leaves often crowded; mostly undershrubs *Thymelaeaceae*
Petals valvate; mostly climbers with swollen nodes and often leaf-opposed inflorescences:
Fruits not enclosed by the calyx; ovary 6-2-locular; fruits baccate, often juicy *Vitaceae*
Fruits completely enclosed by the enlarged calyx; ovary 1-locular; fruit drupaceous *Erythropalaceae*
Calyx-lobes or sepals valvate or calyx much reduced or open in bud:
Disk absent from the flowers (to p. 56):
Trees and shrubs or rarely herbs; flowers not scapose:
Leaves stipulate; stamens hypogynous:
Ovary 8-locular, deeply and laterally lobed; stamens 8, opposite the

petals; shrub with long and short shoots, the leaves clustered on the short shoots; stipules subulate; Socotra *Dirachmaceae*

Ovary and other characters not as above *Sterculiaceae*

Ovary 1-locular with a free central placenta, the ovules pendulous from the top of the ovary; fertile stamens opposite the petals, with barren linear staminodes like spider's legs between them *Medusandraceae*

Leaves without stipules; stamens perigynous *Lythraceae*

Herbs mostly with rosettes of leaves and often lepidote indumentum; flowers in scapose inflorescences; calyx-tube ribbed *Plumbaginaceae*

Tiny herb with axillary flowers; petals 3 *Primulaceae*

Disk present (from p. 55):

Leaves mostly stipulate, stipules sometimes spiny; ovary 2-4-locular; ovules erect; seeds mostly with copious endosperm and large straight embryo; ovary sometimes immersed in the disk *Rhamnaceae*

Leaves without stipules:

Ovules in the whole ovary more than 1; ovary sometimes immersed in the disk *Olacaceae*

Ovule in the whole ovary 1; ovary not immersed in the disk. *Opiliaceae*

Leaves gland-dotted (pellucid punctate) or marked with schizogenous lines (from p. 55):

Leaves simple, without stipules; ovules numerous; trees or shrubs; no tendrils:

Petals hypogynous; ovary 1-locular; ovules numerous on a free basal placenta *Myrsinaceae*

Petals and stamens perigynous; ovary 2- or 3-locular; ovules numerous on an axile placenta; style single with a capitate stigma *Heteropyxidaceae*

Leaves simple, subopposite, with stipules; small trees or shrubs *Rhamnaceae* (*Karwinskia*)

Leaves mostly compound, usually stipulate; inflorescence leaf-opposed; ovules 1 or 2 in each loculus; mostly climbers with tendrils *Vitaceae*

'SYNCARPAE'. **'Axiles'.** Group 18. *Gynoecium composed of* 1 *carpel or of* 2 *or more united carpels with free or united styles, or if carpels free below, then the styles or stigmas united; ovules attached to the central axis or to the base or apex of the ovary; ovary superior, rarely partly immersed in the disk; petals present, free; corolla zygomorphic.*

Petals one fewer than the sepals, often with a scale within; stamens often unilateral or excentric; leaves compound; fruits sometimes samaroid or capsule winged; some climbers *Sapindaceae*
Petals as many as sepals; stamens usually not excentric:
Stamens more than 12:
Stamens and petals hypogynous *Tamaricaceae*
Stamens and petals perigynous or epigynous *Rosaceae*
Stamens 12 or fewer:
Stigmas commissural, sessile; stipules absent; Chile *Francoaceae*
Stigmas or style-arms dorsal to the carpels (i.e. opposite the loculi) or style 1:
Large trees with opposite digitately compound leaves; ovary 3-locular or by abortion 2-1-locular *Hippocastanaceae*
Leaves alternate or radical:
Posterior sepal neither spurred nor saccate; sepals 5, the inner 2 larger and often petaloid, wing-like; anthers opening by an apical pore or rarely by slits; seeds mostly with endosperm:
Ovary 2- or more-locular; fruits not bristly; stipules absent; disk none; leaves simple *Polygalaceae*
Ovary 1-locular; fruits bristly; stipules absent; disk none; leaves simple *Krameriaceae*
Posterior sepal not spurred; anthers opening by slits lengthwise; seeds mostly with copious endosperm:
Sepals more or less free, imbricate *Saxifragaceae*
Sepals united into a tube, valvate *Lythraceae*
Posterior sepal subsaccate:
Stipules large and intrapetiolar; stamens 4, inserted within the unilateral disk *Melianthaceae*
Stipules small and caducous; stamens 3-12, unilateral, partly united *Trigoniaceae*
Stipules very small; stamen 1, fertile; filaments free *Vochysiaceae*
Posterior sepal more or less elongated into a spur:
Spur free and not enclosed (not adnate to) by the pedicel:
Sepals usually 3, rarely 5, often coloured; lateral petals more or less united; stipules absent; leaves not peltate, pinnately nerved *Balsaminaceae*
Sepals 5; lateral petals free; stipules rarely present; leaves peltate or digitately angled lobed or dissected and nerved; upper 2 petals exterior; ovule 1 in each loculus *Tropaeolaceae*
Spur adnate to and concealed within the pedicel; sepals 5, not coloured; petals free; stipules present *Geraniaceae* (*Pelargonium*)

'SYNCARPAE'. **'Axiles'.** Group 19. *Gynoecium composed of* 1 *carpel or* 2 *or more united carpels with free or united styles, or if carpels free below, then the styles or stigmas united; ovules attached to the central axis or to the base or apex of the ovary; ovary superior, rarely partly immersed in the disk; petals present, free; stamens united into* 2 *or more separate bundles.*

- **Leaves** alternate (or if also opposite then often stipulate), or all radical:
  - **Sepals** imbricate:
    - **Trees** or shrubs:
      - **Petals** and stamens hypogynous:
        - **Leaves** neither very small nor fleshy:
          - **Leaves** not glandular-punctate:
            - **Petals** contorted; anthers more or less versatile, sometimes with a gland at the apex *Bonnetiaceae*
            - **Petals** imbricate; anthers never glandular *Theaceae*
          - **Leaves** glandular-punctate, often compound or unifoliolate *Rutaceae*
        - **Leaves** often very small and crowded or fleshy *Tamaricaceae*
      - **Petals** and stamens perigynous:
        - **Leaves** not gland-dotted *Flacourtiaceae*
        - **Leaves** gland-dotted *Myrtaceae*
    - **Herbs** or very small undershrubs:
      - **Stamens** hypogynous:
        - **Carpels** 5, completely united *Geraniaceae*
        - **Carpels** 3, only partially united *Dilleniaceae*
      - **Stamens** more or less perigynous; leaves often very fleshy *Aizoaceae* (*Ficoidaceae*)
  - **Sepals** valvate; hairs on the leaves etc. often stellate:
    - **Stamens** either monadelphous or if in bundles then some sterile *Sterculiaceae*
    - **Stamens** in separate bundles and all fertile *Tiliaceae*
- **Leaves** opposite, often gland-dotted or with resinous lines:
  - **Stamens** hypogynous:
    - **Styles** free from the base or nearly so; flowers bisexual; shrubs or herbs; cotyledons well-developed *Hypericaceae*
    - **Styles** mostly more or less united or stigma one and sessile or subsessile; flowers mostly unisexual; trees or shrubs often with numerous close lateral nerves; cotyledons usually minute or obsolete *Clusiaceae*
  - **Stamens** more or less perigynous; petals sometimes numerous; leaves often very thick and fleshy *Aizoaceae* (*Ficoidaceae*)

'SYNCARPAE'. **'Axiles'.** Group 20. *Gynoecium composed of* 1 *carpel or* 2 *or more united carpels with free or united styles, or if carpels free below then the styles or stigmas united; ovules attached to the central axis or to the base or apex of the ovary; ovary superior, rarely partly immersed in the disk; petals present, more or less united into a tube (sympetalous); stamens more than twice as many as the corolla-lobes; corolla actinomorphic.*

Anthers opening by apical pores or short pore-like slits:
- **Corolla-tube** very short; stamens very numerous, adnate at the base to the corolla; anthers not tailed — *Saurauiaceae*
- **Corolla-tube** usually fairly long; stamens rarely very many more than double the number of the corolla-lobes, not adnate to the tube; anthers often with tails — *Ericaceae*

Anthers opening by longitudinal slits:
- **Flowers** bisexual:
  - **Leaves** simple but sometimes deeply divided:
    - **Bracts** more or less pouch-like and adnate to the pedicels; stems climbing or plants epiphytic — *Marcgraviaceae*
    - **Bracts** and habit not as above:
      - **Stamens** hypogynous:
        - **Sepals** imbricate:
          - **Trees** or shrubs — *Theaceae*
          - **Herbs** or slightly woody at the base — *Molluginaceae*
        - **Sepals** valvate:
          - **Stamens** free or nearly so; anthers 2-locular — *Tiliaceae*
          - **Stamens** monadelphous; anthers 1-locular — *Malvaceae*
      - **Stamens** perigynous or epigynous; sepals usually imbricate:
        - **Leaves** often gland-dotted and opposite — *Myrtaceae*
        - **Leaves** not gland-dotted, alternate — *Lecythidaceae*
  - **Leaves** digitately compound, opposite or alternate — *Caryocaraceae*
- **Flowers** unisexual:
  - **Disk** absent; leaves without stipules — *Ebenaceae*
  - **Disk** present; leaves usually with stipules — *Euphorbiaceae*

'SYNCARPAE'. **'Axiles'.** Group 21. *Gynoecium composed of* 1 *carpel or* 2 *or more united carpels with free or united styles, or if carpels free below, then the styles or stigmas united; ovules attached to the central axis or to the base or apex of the ovary; ovary superior, rarely partly immersed in the disk; petals present, more or less united into a tube (sympetalous); stamens as many as the corolla-lobes and alternate with them or twice as many or fewer; corolla actinomorphic ('regular'); leaves alternate or all radical or reduced to scales, or rarely absent.*

Leafless nevergreen parasites destitute of chlorophyll; flowers bisexual:
Ovary 1-4-6-lobed and locular; ovules very numerous in each loculus; seeds minute with undivided embryo *Monotropaceae*
Ovary 10-14-lobed and about 20-28-locular; ovule solitary in each loculus owing to spurious division of the carpels; fruit at length opening by an irregular circumscissile slit *Lennoaceae*
Ovary more or less 2-locular; ovules 4 in each ovary; stems thread-like and often forming a mass on other plants; corolla-tube with scales alternating with the lobes *Cuscutaceae*
Leaves more or less green and normally developed; not parasites or rarely so:
Leaves stipulate, stipules sometimes soon falling off, sometimes foliaceous:
Leaves densely covered with viscid gland-tipped processes, mostly all radical; often stemless herbs; flowers usually in simple circinate cymes; placentas sub-basal; styles 3-5, mostly free *Droseraceae*
Leaves etc. not as above:
Trees or shrubs the latter sometimes straggling but not by hooks or tendrils:
Petals bifid or bilobed, not contorted; flowers bisexual or unisexual; calyx not wing-like in fruit *Dichapetalaceae*
Petals not bilobed: *(Chailletiaceae)*
Flowers bisexual; petals contorted; calyx wing-like and enlarged in fruit *Dipterocarpaceae*
Flowers unisexual; petals not contorted; ovary usually 3-lobed and 3-locular; calyx not wing-like in fruit *Euphorbiaceae*
Climbing shrubs; petals contorted:
Climbing by means of hooked branches; petals only slightly connate; anthers basifixed *Ancistrocladaceae*
Climbing by means of tendrils; petals connate into a tube; anthers dorsifixed and versatile; leaves pinnate *Cobaeaceae*
Leaves without stipules:
Stamens free from the corolla or only slightly adnate to its base (hypogynous or perigynous) (to p. 63):
Anthers opening by terminal pores (to p. 61):
Anthers often with appendages; calyx persistent, sometimes petaloid; corolla-lobes contorted or imbricate; woody often 'healthy' plants or large shrubs or small trees (Rhododendrons) *Ericaceae*
Anthers without appendages:

Stamens numerous:
Flowers bisexual; leaves with numerous strong parallel nerves diverging from the midrib; sepals very imbricate *Saurauiaceae*
Flowers bisexual; leaves with few lateral nerves; sepals valvate *Tiliaceae* (*Antholoma*)
Flowers dioecious; leaves pinnately nerved *Ebenaceae*
Stamens 5-10:
Ovary 3-2-locular; stamens 5; leaves linear, glandular-pilose; pollen-grains not united in tetrads *Byblidaceae*
Ovary perfectly or imperfectly 5-locular; stamens 10-8; leaves not as above; pollen-grains united in tetrads *Pyrolaceae*
Anthers opening by slits lengthwise or rarely transversely (from p. 60):
Stamens 4-6, rarely fewer (rarely united into a tube) (to p. 62):
Leaves gland-dotted; ovary mostly deeply lobed; petals connivent mostly by their claws; disk usually conspicuous between the stamens and ovary *Rutaceae*
Leaves not gland-dotted; ovary mostly not deeply lobed:
Petals only shortly united at the base:
Disk present in the flowers, usually conspicuous but not adherent to the ovary:
Corolla-segments valvate; ovules 2-5 or rarely in each loculus:
Stamens opposite the petals or more numerous than them *Olacaceae*
Stamens the same number as and alternate with the petals *Icacinaceae*
Corolla-segments or lobes imbricate; ovules numerous:
Leaves not small and not ericoid *Pittosporaceae*
Leaves small and ericoid *Tamaricaceae*
Disk absent, or if present, adherent to the ovary:
Leaves simple or 1-foliolate:
Stamens with free or nearly free filaments:
Ovules numerous:
Epiphytic or scandent shrubs; bracts 3-lobed *Marcgraviaceae*
Not epiphytic; bracts not 3-lobed *Pittosporaceae*
Ovules 1 or 2 or very few in each loculus:
Calyx-lobes imbricate; indumentum when present not stellate *Aquifoliaceae*
Calyx-lobes valvate or open; indumentum often stellate or lepidote *Styracaceae*
Calyx mostly quite reduced; indumentum when present not as above *Icacinaceae*
Stamens with filaments united into a cylindric tube *Meliaceae*
Leaves pinnate; ovules collateral, 2 in each loculus *Connaraceae*
Petals united high up but sometimes free at the base:
Calyx completely reduced; male flowers in dense heads; climbers; stamens 3 *Icacinaceae*
Calyx well developed:
Calyx-lobes valvate, herbaceous; anthers connivent around the style:

Herbs, not aquatic, often with dull or black hairs; flowers blue *Campanulaceae (Cyananthus)*
Trees or shrubs *Burseraceae*
Climbers *Styracaceae*
Calyx-lobes imbricate:
Anthers 1-locular, opening by a single slit lengthwise; shrubs or shrublets; flowers bisexual *Epacridaceae*
Anthers 2-locular:
Herbs from a woody rhizome; leaves linear or spathulate *Stackhousiaceae*
Aquatic or terrestrial herbs; flowers often polygamous or unisexual; calcyx-lobes persistent in fruit *Plantaginaceae*
Shrublets; flowers bisexual, solitary *Prionotaceae**
Stamens more than 6 (from p. 61):
Stamens connate into a distinct tube *Meliaceae*
Stamens free or slightly connate only at the base:
Leaves gland-dotted, often compound or unifoliolate, sometimes lepidote *Rutaceae*
Leaves not gland-dotted, simple or very rarely compound:
Disk absent; corolla sometimes long and tubular:
Flowers bisexual; calyx very much imbricate *Fouquieriaceae*
Flowers unisexual; calyx not or very slightly imbricate *Ebenaceae*
Disk present, or if inconspicuous then the corolla-tube very short:
Leaves well developed, not small and not ericoid:
Leaves simple; bark not bitter:
Petals contorted or imbricate; ovules pendulous from the top of the axis *Cyrillaceae*
Petals imbricate; ovules axile and spreading *Ericaceae*
Petals valvate *Olacaceae*
Leaves compound or unifoliolate; bark usually bitter *Simaroubaceae*
Leaves very small and ericoid; flowers in spikes or racemes *Tamaricaceae*
Disk absent; corolla-tube usually very short:
Trees and shrubs (to p. 63):
Ovules 1 or 2 in each loculus; indumentum not stellate; calyx-lobes imbricate:
Calyx-tube very short:
Leaves digitately foliolate; petals connate into a calyptra *Caryocaraceae*
Leaves simple; petals not calyptrate *Aquifoliaceae*
Calyx-tube long and slender; leaves simple *Thymelaeaceae*
Ovules few; indumentum often stellate or lepidote (to p. 63):
Calyx-lobes valvate or open in bud (to p. 63):
Calyx more or less adnate to the ovary; style slender; ovules spreading from the axis of the ovary *Styracaceae*
Calyx not adnate to the ovary; style very short; ovules pendulous from the top of the axis *Cyrillaceae*

Calyx lobes or sepals imbricate *Theaceae*
Ovules numerous in each loculus (from p. 62):
Sepals valvate; petals with a scale-like contracted base; disk thick and fleshy, broadly turbinate *Pentadiplandraceae*
Sepals imbricate; petals not scale-like at the base; disk absent *Theaceae*
Herbs; sepals often scarious *Molluginaceae*
Stamens inserted on the corolla-tube, rarely in a column adnate to the stigma (from p. 60):
Style gynobasic, arising from between the lobes of the ovary:
Style single; fruit composed of pyrenes or nutlets:
Flowers axillary; corolla-lobes plicate; ovary 5-locular; ovules and seeds in separate locelli of the ovary and fruits respectively *Nolanaceae*
Flowers mostly in scorpioid cymes; corolla-lobes contorted or imbricate; ovary 4-2-locular; plants often very scabrid or hispid *Boraginaceae*
Styles 2; fruits deeply 2-lobed, lobes nut-like *Convolvulaceae*
Style or styles not gynobasic, terminal on the ovary:
Stamens as many as or more than the corolla-lobes (to foot of p. 64):
Corolla-lobes contorted in bud; ovary 1-3-locular:
Carpels completely united:
Ovary mostly 3-locular; capsule loculicidae *Polemoniaceae*
Ovary 1- or 2-locular; style at most once-lobed:
Style single, undivided; embryo cyclic, spiral or straight in thin or no endosperm *Solanaceae*
Styles 2 or style deeply 2-lobed; embryo very small in copious endosperm *Hydrophyllaceae*
Ovary 4-locular; styles twice 2-lobed *Ehretiaceae*
Carpels more or less free, especially below and more so in fruit:
Pollen not agglutinated into wax-like masses; corona absent *Apocynaceae*
Pollen agglutinated into wax-like masses; corona present *Asclepiadaceae*
Corolla-lobes imbricate in bud (to p. 64):
Flowers crowded in dense leafy heads; shrubs with sessile imbricate leaves; ovary 2-locular; South Africa *Bruniaceae*
Flowers in dense spikes; herbs with radical leaves and broad sheathing petioles *Plantaginaceae*
Flowers not in heads, rarely in spikes, sometimes corymbose or cymose:
Ovary 3-10-locular (to p. 64):
Stamens numerous, several seriate; racemes or spikes axillary *Symplocaceae*
Stamens 15 or more, sometimes in bundles of three alternating with the corolla-lobes; anthers banfixed or versatile *Theaceae*
Stamens twice as many as corolla-lobes, not in threes; staminodes usually present; ovule 1 in each loculus *Sapotaceae*
Stamens 10 or 5:
Styles twice 2-lobed; trees and shrubs *Ehretiaceae*
Styles only once 2-lobed:

Anthers 2-locular or if 1-locular then stamens connate with as many alternate staminodes:
Flowers not cymose *Diapensiaceae*
Flowers cymose *Boraginaceae*
Anthers 1-locular, opening by a single slit; no staminodes *Epacridaceae*
Ovary 1- or 2-locular or rarely spuriously 4-locular (from p. 63):
Trees, shrubs or climbing herbs; fruit a pyrene or nut:
Fruits winged; stigmas 2, unequal, one elongated and columnar and persistent, the other short, capitate and stipitate *Cardiopteridaceae*
Fruits not winged; stigmas equal, clavate or capitate:
Style mostly twice 2-lobed *Ehretiaceae*
Style once lobed or fid *Chloanthaceae*
Herbs:
Style deeply 2-lobed or styles 2; all the leaves alternate or radical:
Fruits indehiscent *Ehretiaceae*
Fruits capsular *Hydrophyllaceae*
Style undivided or shortly lobed:
Lower leaves opposite *Gentianaceae*
All the leaves alternate:
Ovary-loculi placed ad-and-ab-axial *Scrophulariaceae*
Ovary-loculi placed oblique to the axis *Solanaceae*
Corolla-lobes valvate or plicate in bud, in the latter case the limb may be somewhat twisted but the lobes not really overlapping (from p. 63):
Ovules numerous in each ovary-loculus:
Flowers without a corona; pollen granular:
Trees with axillary flowers; stamens about 30, often in pairs *Ebenaceae*
Mostly herbs with a definite number (few) of stamens:
Ovary partly inferior *Campanulaceae*
Ovary superior *Solanaceae*
Flowers with a corona; pollen agglutinated in wax-like masses *Asclepiadaceae*
Ovules 1-4 in each ovary-loculus:
Ovules basal, erect:
Anthers not united; widely distributed families:
Ovules and seeds not separated into separate locelli of the ovary and fruit *Convolvulaceae*
Ovules and seeds separated into separate locelli in the ovary and fruit *Nolanaceae*
Anthers united into a tube around the style; Australia *Brunoniaceae*
Ovules pendulous from the top of the ovary:
Fruits drupaceous; trees or shrubs; see also *Boraginaceae* *Icacinaceae*
Fruits baccate or capsular *Solanaceae*
Ovules pendulous from the top of a basal placenta *Olacaceae*
Stamens fewer than the corolla-lobes; leaves mostly opposite (from p. 63):

Stamens more than 1:
Ovules numerous; corolla not dry and scarious (to p. 65):
Ovary completely 2-locular:
Herbs:
Corolla-lobes in duplicate or contorted-plicate; ovary-loculi placed oblique to the axis *Solanaceae*
Corolla-lobes imbricate; ovary-loculi placed ad- and ab-axial *Scrophulariaceae*
Shrubs, erect or climbing; stamens 2; leaves simple or compound *Oleaceae*
Ovary 1-locular or imperfectly 2-locular by the intrusive placentas *Gesneriaceae*
Ovules 1-8 in each loculus of the ovary; capsule circumscissile; corolla scarious, with 4 lobes; mostly herbs with radical leaves and spicate inflorescences *Plantaginaceae*
Ovules 1 or 2 in each loculus of the ovary; corolla not scarious, and other characters not combined:
Corolla-lobes not more than the calyx-lobes:
Corolla-lobes quite equal; ovules mostly pendulous:
Shrublets of ericoid habit and very small leaves; anthers 1-locular *Epacridaceae*
Habit not as above; anthers 2-locular *Oleaceae*
Corolla-lobes slightly unequal or limb oblique:
Ovules erect *Verbenaceae*
Ovules pendulous *Myoporaceae*
Corolla-lobes about 3 times as many as the calyx-lobes *Sapotaceae*
Ovules few; erect or scandent shrubs sometimes with pinnate leaves; stamens 2; corolla-lobes imbricate *Oleaceae*
Stamen 1:
Trees; corolla articulating above the base; flowers in corymbose panicles, bisexual *Antoniaceae*
Herbs, stemless; flowers spicate-capitate, polygamous; ovary 1-locular, 1-ovuled *Plantaginaceae*

'SYNCARPAE'. **'Axiles'.** Group 22. *Gynoecium composed of* 1 *carpel or* 2 *or more united carpels with free or united styles, or if carpels free below, then the styles or stigmas united; ovules attached to the central axis, or the adaxial suture (when carpels free) or to the base or apex of the ovary; ovary superior, rarely partly immersed in the disk; petals present, more or less united into a tube (sympetalous); stamens if the same number as the corolla-lobes then alternate with them, or up to twice as many or fewer; corolla actinomorphic ('regular'); leaves opposite or verticillate.*

Anthers opening by apical pores or short pore-like slits:
- **Filaments** of the stamens often geniculate and inflexed in bud; anthers opening by a single pore (rarely 2 pores); leaves mostly with the main nerves longitudinally parallel with the midrib and margin *Melastomataceae*
- **Filaments** of the stamens not as above; anthers opening by 2 pores; leaves without longitudinal parallel nerves:
  - **Woody** plants with usually evergreen leaves and these often very small ('ericoid'); stamens not inserted on the corolla (hypogynous) *Ericaceae*
  - **Herbs:** stamens inserted on the corolla (epipetalous) *Gentianaceae*

Anthers opening by slits lengthwise:
- **Leaves** gland-dotted, glandular or pustulate, or glandular-serrate; petals usually only shortly united:
  - **Stamens** not inserted on the corolla:
    - **Petals** and stamens hypogynous; leaves pellucid-punctate *Rutaceae*
    - **Petals** and stamens perigynous; leaves glandular-serrate *Escalloniaceae*
  - **Stamens** inserted on the corolla (epipetalous):
    - **Style** or style terminal *Loganiaceae*
    - **Style** inserted at the inner base of the vertically lobed ovary (gynobasic):
      - **Stamens** not didynamous *Boraginaceae*
      - **Stamens** 4, didynamous *Lamiaceae* (*Labiatae*)
- **Leaves** not gland-dotted; petals usually joined high up:
  - **Stamens** double the number of the corolla-lobes; petals united only at the base:
    - **Flowers** bisexual; mostly herbs:
      - **Corolla-lobes** not numerous; anthers inflexed in bud *Lythraceae*
      - **Corolla-lobes** numerous; anthers erect in bud *Molluginaceae*
    - **Flowers** dioecious; trees or shrubs; anthers erect in bud *Ebenaceae*
  - **Stamens** the same number as or fewer than the corolla-lobes:
    - **Style** gynobasic; ovary deeply and vertically 4-lobed *Boraginaceae*
    - **Style** or styles or stigma terminal:
      - **Anthers** 1-locular and opening by a single slit; stamens sometimes hypogynous *Epacridaceae*
      - **Anthers** 2-locular, opening by 2 slits:
        - **Stamens** hypogynous, more or less inserted on a disk and free from the corolla (to top of p. 67) *Ericaceae*

Stamens inserted on the corolla-tube:
Corolla-lobes contorted (see also p. 68):
Pollen grains granular and not agglutinated into masses; corona none:
Carpels separate but styles or stigmas united *Apocynaceae*
Carpels united into an ovary:
Shrubs or trees *Potaliaceae*
Herbs *Polemoniaceae*
Pollen agglutinated into waxy masses; seeds crowned by a tuft of hairs *Asclepiadaceae*
Corolla-lobes imbricate or plicate (see also p. 68):
Pollen agglutinated into waxy masses; corona usually present *Asclepiadaceae*
Pollen granular:
Style single, entire or very shortly fid, often with a large more or less capitate stigma:
Carpels 2, free, but more or less united above by their styles or stigmas:
Stamens without a coronal appendage; pollen granular and transported freely; fruits sometimes indehiscent *Apocynaceae*
Stamens with a coronal appendage; pollen transported by spathulate carriers; fruits follicular, dehiscing ventrally *Periplocaceae*
Carpels completely united below the style or styles:
Stamens more than 2:
Corolla scarious; mostly herbs with radical leaves and dense spikes of flowers *Plantaginaceae*
Corolla not scarious; other characters not as above:
Trees, shrubs or shrublets:
Ovules 1 or 2, erect from the base of the ovary; leaves verticillate *Stilbaceae*
Ovules more than 2, spreading from the central axis:
Indumentum when present neither glandular, stellate nor lepidote; intraxylary phloem present *Loganiaceae*
Indumentum when present glandular-stellate or lepidote; intraxylary phloem absent *Buddleiaceae*
Herbs or scramblers; leaves not stipulate *Solanaceae*
Stamens 2; leaves simple or pinnate *Oleaceae*
Styles separate or with more than 1 separate stigmas:
Trees or shrubs:
Ovules numerous *Loganiaceae*
Ovules 1 or 2 in each loculus:
Rudimentary stipules often present; stamens and corolla-lobes 4 each *Salvadoraceae*
Stipules absent:
Seeds without endosperm *Verbenaceae*
Seeds with endosperm *Chloanthaceae*
Herbs or herbaceous climbers:
Corolla-lobes mostly plicate; fruits mostly indehiscent and baccate; embryo much curved, annular or spirally coiled *Solanaceae*

Corolla lobes imbricate; fruits mostly capsular; embryo small and straight *Gentianaceae* and *Hydrophyllaceae*

Corolla-lobes valvate:

Carpels united their full length; pollen granular, not in waxy masses:

Leaves verticillate, linear; seeds not winged *Retziaceae**

Leaves not verticillate:

Stipules intrapetiolar *Rubiaceae* (*Gaertnera*)

Stipules absent or when present not intrapetiolar:

Stamens 5, 4 rarely 1:

Fruit a septicidal capsule; seeds winged at each end or all around; stamens sometimes only 1; stamens inserted in the throat of thecorolla *Antoniaceae*

Fruit as above but seeds not winged; stamens inserted in the corolla-tube *Spigeliaceae*

Fruits indehiscent, drupaceous or baccate; seeds not winged; branches sometimes spiny; leaves often 3-nerved from the base *Strychnaceae*

Stamens 2 *Oleaceae*

Carpels free except for the styles or stigmas; pollen agglutinated into waxy masses; corona often present *Asclepiadaceae*

'SYNCARPAE'. **'Axiles'.** Group 23. *Gynoecium composed of* 1 *carpel or* 2 *or more united carpels with free or united styles, of if carpels free below, then the styles or stigmas united; ovules attached to the central axis or to the base or apex of the ovary; ovary superior, rarely partly immersed in the disk; petals present, more or less united into a tube (sympetalous); stamens the same number as and opposite to the corolla-lobes.*

Ovule solitary in the whole ovary or in each loculus of the ovary; style often lobed :
Trees or shrubs, often with very hard wood:
Flowers mostly dioecious; petals imbricate or valvate; stamens mostly free from or attached only to the base of the corolla *Ebenaceae*
Flowers bisexual; stamens inserted on the corolla (epipetalous):
Corolla-lobes imbricate:
Leaves not sheathing at the base:
Flowers fasciculate in the leaf-axils or at the older nodes; leaves alternate, rarely opposite; stipules usually absent; staminodes often present *Sapotaceae*
Flowers in racemes or panicles; leaves opposite or subopposite, often with large pits in the axils of the nerves; stipules small, mostly caducous; staminodes 5, alternate with the stamens *Sarcospermataceae*
Leaves sheathing at the base *Plumbaginaceae*
Corolla-lobes valvate:
Inflorescence not leaf-opposed; leaves simple:
Ovules more than 1 in the whole ovary *Olacaceae*
Ovule 1 in the whole ovary *Opiliaceae*
Inflorescence leaf-opposed; leaves often compound *Vitaceae*
Herbs or subherbaceous climbers:
Corolla-lobes valvate; tendrils often present; inflorescence usually cymose-paniculate, leaf-opposed; leaves mostly with stipules adnate to the petiole *Vitaceae*
Corolla-lobes imbricate:
Stamens connate into a column, free from the corolla-lobes; stipules absent *Menispermaceae*
Stamens free from each other and from the corolla; stipules scarious, rarely absent *Portulacaceae*
Stamens more or less adnate to or inserted on the corolla *Plumbaginaceae*
Ovules 2 or more in each loculus or in the whole ovary; style mostly undivided; placentas often free-basal:
Trees or shrubs, often gland-dotted leaves or with schizogenous lines (to p. 70):
Anthers not transversely septate; seeds with endosperm and short embryo; cotyledons not connate:
Anthers extrorse, mostly with a produced connective, opening by slits lengthwise; staminodes present, alternating with the corolla-lobes *Theophrastaceae*

Anthers introrse, opening lengthwise or by apical pores; no staminodes *Myrsinaceae*

Anthers transversely septate; seed solitary, without endosperm, elongated; embryo long and arcuate, germinating in the fruit; cotyledons connate into a tube *Aegicerataceae*

Herbs or climbing shrubs usually with compound often pellucid-punctate leaves and leaf-opposed inflorescences *Vitaceae*

Herbs, often with rosettes of leaves; leaves not gland-dotted:

Corolla not split on one side, never calyptriform *Primulaceae*

Corolla split on one side or calyptrate; annual herbs *Portulacaceae*

'SYNCARPAE'. **'Axiles'.** Group 24. *Gynoecium composed of* 1 *carpel or* 2 *or more united carpels with free or united styles, or if carpels free below, then the styles or stigmas united; ovules attached to the central axis or to the base or apex of the ovary; ovary superior, rarely partly immersed in the disk; petals present, more or less united into a tube (sympetalous); corolla zygomorphic ('irregular').*

Stamens as many as or more than the corolla-lobes (rarely some sterile or without anthers):
 Ovary vertically deeply 4-lobed; style gynobasic:
  Leaves alternate, often clothed with bulb-like-based hairs *Boraginaceae*
  Leaves opposite *Lamiaceae (Labiatae)*
 Ovary not vertically lobed or only slightly so; style not gynobasic, terminal or subterminal:
  Lower sepal elongated into a tubular spur; ovary 5-locular *Balsaminaceae*
  Lower sepal not spurred; ovary usually 2-locular:
   Ovary 1-locular with a basal placenta and numerous ovules; insectiverous marsh, aquatic or rarely epiphytic herbs; fertile stamens 2; lower lip of corolla saccate or spurred *Lentibulariaceae*
   Ovary 2- or more-locular; other characters not as above:
    Ovules numerous:
     Leaves simple:
      Stamens inserted on the corolla-tube; anthers rarely opening by pores:
       Corolla-lodes induplicate-valvate or contorted:
        Corolla mostly subactinomorphic; stigma not indusiate *Solanaceae*
        Corolla zygomorphic, the tube sometimes split on one side; stigma indusiate *Goodeniaceae*
       Corolla-lobes imbricate or folded (plicate) *Scrophulariaceae*
      Stamens free from the corolla-tube; anthers opening by pores *Ericaceae ('Azalea')*
     Leaves compound or 1-foliolate *Bignoniaceae*
    Ovules few:
     Anthers 2-locular, opening by a longitudinal slit:
      Stamens alternate with or more than the corolla-lobes; mostly woody plants:
       Leaves not glandular-punctate:
        Seeds without endosperm; ovules 1 or 2 in each loculus, erect or rarely pendulous *Verbenaceae*
        Seeds with endosperm; ovules 2 in each loculus *Chloanthaceae*
       Leaves glandular-punctate *Rutaceae*
      Stamens the same number and opposite the corolla-lobes; small herbs *Primulaceae*
     Anthers 1-locular, opening by a longitudinal slit *Selaginaceae*
     Anthers 1- or 2-locular, opening by terminal pore *Polygalaceae*
Stamens fewer than the corolla-lobes; leaves mostly opposite:
 Placenta free-basal, more or less globose; ovules numerous; ovary 1-locular;

fertile stamens 2; stigma sessile or style very short; insectivorous aquatic or marsh, rarely epiphytic herbs *Lentibulariaceae*

Placentas axile, or if subbasal ovules few:

Ovules numerous in the whole ovary or in each loculus, or if only 2, then superposed:

Anthers mostly free from each other; ovary completely 2-locular with the placentas on the septa:

Loculi of the ovary placed ad- and ab-axial (i.e. in line with the axis) *Scrophulariaceae*

Loculi of the ovary oblique to the axis *Salpiglossidaceae**

Anthers often coherent; ovary imperfectly 2-locular by the variously intrusive parietal placentas, the latter placed right and left of the floral axis *Gesneriaceae*

Anthers free but often more or less connivent; ovary mostly 2-locular and then the loculi ad- and abaxial (antero-posterior); leaves often compound and usually ending in a tendril; woody plants, very rarely herbaceous; seeds often winged, transverse to the length of the fruit *Bignoniaceae*

Anthers often connivent in pairs; ovary 1-4-locular; fruit often with a hard endocarp; seeds not winged:

Herbs or rarely shrubs with vescicular glands; leaves opposite or the upper alternate; fruits sometimes prickly or spiny; disk mostly inconspicuous *Pedaliaceae*

Herbs or rarely shrubs with tumid nodes and opposite or verticillate leaves often prominently marked with cytoliths; seeds inserted on hardened outgrowths from the central axis (retinacula); valves of the capsule opening elastically from the apex; disk cupular or annular *Acanthaceae*

Shrubs; nodes not tumid; leaves alternate; 2 inner lobes on the corolla broadly emarginate or bifid; fruit a drupe; disk subannular or lobed *Dichapetalaceae* (*Chailletiaceae*)

Ovule solitary in each loculus of the ovary, or if 2 then collateral:

Leaves alternate or radical:

Filaments free from each other:

Anthers 2-locular; leaves usually studded with resinous glands *Myoporaceae*

Anthers 1-locular, opening by a slit; leaves not studded with glands:

Ovary 2-locular *Selaginaceae*

Ovary 1-locular; inflorescence girt by a common involucre *Globulariaceae*

Filaments more or less monadelphous and the tube split along the top; anthers 1- or rarely 2-locular, opening by a terminal pore:

Ovary 2- or more-locular; fruits not bristly *Polygalaceae*

Ovary 1-locular; fruits bristly *Krameriaceae*

Leaves opposite or verticillate:

Ovary deeply and vertically 4-lobed; style gynobasic *Lamiaceae* (*Labiatae*)

Ovary entire, not vertically lobed or only slightly so; style terminal:

Filaments forming a sheath split on its upper side *Polygalaceae*
Filaments free from each other or connate in pairs:
Calyx not bilabiate; ovary 9-2-locular:
Ovules collateral in the loculi; fruit mostly of 2 or 4 pyrenes:
Seeds without endosperm; leaves not verticillate *Verbenaceae*
Seeds with endosperm; leaves verticillate:
Inflorescence spicate; South Africa *Stilbaceae*
Inflorescence mostly cymose; Australia and neighbouring islands *Chloanthaceae*
Ovules superposed in the loculi; fruit a capsule or drupe, often armed *Pedaliaceae*
Calyx bilabiate; ovary 1-locular with 1 erect ovule; 3 calyx-lobes hooked *Phrymaceae*

'SYNCARPAE'. **'Axiles'.** Group 25. *Gynoecium composed of* 1 *carpel or* 2 *or more united carpels with free or united styles, or if carpels free below, then the styles or stigmas united; ovules attached to the central axis or to the base or apex of the ovary; ovary superior, rarely partly immersed in the disk; petals absent; bisexual or male flowers* (*and often the female*) *with a distinct calyx; leaves alternate or radical or reduced to scales.*

Parasitic plants with the leaves reduced to scales:
Anthers opening by slits lengthwise:
Flowers in spikes; sepals imbricate; seeds numerous, minute, with copious endosperm and undivided embryo *Monotropaceae*
Flowers densely crowded; sepals valvate; seed solitary *Balanophoraceae*
Anthers opening by valves; twining plants with whip-like stems and branches *Lauraceae*

Not parasitic; leaves normally developed:
Submerged fresh-water herbs often greatly resembling mosses, hepatics and algae; stamens 1-4; ovules numerous on a central placenta *Podostemaceae*

Plants not as described above:
Leaves with stipules, these sometimes partly adnate to the petiole, or minute (to p. 76):
Stamens monadelphous, usually numerous; calyx-lobes or sepal, mostly valvate:
Anthers 2-locular; flowers bisexual:
Sepals or calyx-lobes valvate; indumentum usually stellate *Sterculiaceae*

Sepals or calyx-lobes imbricate:
Carpels connate around a central column often dilated into a flat disk at the top; embryo curved around the central fleshy endosperm *Gyrostemonaceae*
Carpels not arranged as above; embryo straight, no endosperm *Rosaceae*
Anthers 1-locular; flowers bisexual *Malvaceae*
Anthers 2-locular; flowers unisexual:
Disk of female flowers not large and fleshy and not enclosing the gynoecium *Euphorbiaceae*
Disk of female flowers large and fleshy and covering the gynoecium *Scyphostegiaceae*

Stamens free or shortly united only at the base:
Stamens the same number as the sepals or calyx-lobes and alternate with them (to p. 75):
Ovary composed of more than 1 carpel:
Herbs with radical, alternate or pseudo-verticillate leaves, or shrublets; stipules membranous, sometimes sheathing the stem *Molluginaceae*

Trees, shrubs or climbers *Rhamnaceae*
Ovary composed of 1 carpel *Petiveriaceae*
Stamens if the same number as the sepals or calyx-lobes then opposite to them, or more numerous or fewer, filaments sometimes connate at the base (from p. 74):
Leaves compound:
Flowers unisexual:
Ovary 1-locular *Moraceae*
Ovary 2- or more-locular:
Ovules 1 or 2 in each loculus or ovary:
Leaves pinnate *Rhoipteleaceae*
Leaves digitate *Euphorbiaceae*
Ovules numerous in each loculus *Saxifragaceae*
Flowers bisexual:
Flowers actinomorphic ('regular'):
Style basal or ventral (rarely terminal) on the 1-carpelled ovary; seeds without endosperm; stipules often partly adnate to the petiole *Rosaceae*
Style or styles more or less terminal; ovary of more than 1 carpel; seeds usually with endosperm *Saxifragaceae*
Flowers slightly zygomorphic ('irregular'); leaves pinnate; stipules intra-petiolar, often large; disk unilateral; stamens 4 *Melianthaceae*
Leaves simple, though sometimes deeply divided:
Ovary 2- or more-locular (to p. 76):
Flowers unisexual or polygamous:
Ovary quite superior; anthers opening by slits lengthwise:
Male flowers not in catkins:
Embryo of seed straight in the middle of the endosperm *Euphorbiaceae*
Embryo of the seed much curved around the central fleshy endosperm:
Fruit indehiscent; ovary carpels not connate around a central column *Phytolaccaceae*
Fruit dehiscent; ovary-carpels connate around a central column often dilated into a disk-like top *Gyrostemonaceae*
Male flowers in catkins *Betulaceae*
Ovary semi-superior, mostly 2-locular; anthers opening by valves *Hamamelidaceae*
Flowers bisexual:
Trees or shrubs; carpels often free at the top; stipules often paired or adnate to the petiole (to p. 76):
Calyx-lobes or sepals imbricate:
Ovary quite superior; stipule single, adnate to the petiole; tree with long and short shoots; styles at length sub-basal *Tetracentraceae*
Ovary inferior or semi-inferior; stipules paired, free *Hamamelidaceae*

Calyx-lobes or sepals valvate; ovary superior, style terminal *Tiliaceae*

Herbs (from p. 75):

Stamens perigynous; ovary 1-3-locular *Saxifragaceae*

Stamens hypogynous; ovary 3-5-locular, very rarely 2-locular; stipules often lacerate or fimbriate *Molluginaceae*

Ovary 1-locular, mostly of 1 carpel (from p. 75)

Stipules ochreate, i.e. sheathing and more or less membranous around the stem; disk usually present; stigmas capitate, peltate or fimbriate; fruit a nut, indehiscent, sometimes winged *Polygonaceae*

Stipules not as above:

Stamens inflexed in bud; flowers unisexual:

Trees or shrubs, often with milky juice *Moraceae*

Herbs, some with stinging hairs; leaves mostly with prominent cystoliths *Urticaceae*

Stamens erect in bud:

Flowers bisexual; style sometimes basal:

Flowers not in racemes *Rosaceae*

Flowers in racemes *Peteveriaceae*

Flowers unisexual:

Flowers in catkins; seeds with numerous fine hairs arising from the funicle *Salicaceae*

Flowers not in catkins; seeds without fine hairs:

Fruits not winged *Rosaceae*

Fruits often winged or appendaged *Ulmaceae*

Male flowers paniculate *Cannabaceae*

Leaves without stipules (from p. 74):

Stamens the same number as and alternating with the sepals or calyx-lobes, or in the center of the flower:

Leaves compound, mostly pinnate:

Flowers bisexual or polygamo-dioecious, not involucrate *Sapindaceae*

Flowers dioecious, the females solitary, enclosed in and adnate to an involucre *Julianaceae*

Flowers dioecious, paniculate or racemose, the female not in an involucre *Anacardiaceae* (*Pistacia*)

Leaves simple:

Small shrubs with minute leaves *Aizoaceae* (*Ficoidaceae*)

Herbs; leaves not minute:

Stamens 5; lower leaves opposite, upper alternate; ovules numerous on a free basal placenta *Primulaceae* (*Glaux*)

Stamens 1 or 2; stigma sessile; leaves in a rosette, toothed; ovule 1, pendulous *Circaeasteraceae*

Trees or shrubs with unisexual or polygamo-dioecious flowers; ovules 1 or 2, axile *Sapindaceae*

Stamens the same number as the sepals or calyx-lobes and opposite to them, or more numerous or fewer:

Herbs:

Ovary composed of 1 carpel; leaves ternately compound; flowers bisexual, paniculate or racemose, usually very small *Ranunculaceae*

Ovary composed of 3 carpels; leaves tubular; flowers racemose *Sarraceniaceae* (*Heliamphora*)

Ovary composed of 2 carpels; leaves pinnately or digitately lobed *Papaveraceae*

Ovary composed of 5-25 carpels; leaves simple *Phytolaccaceae*

Shrubs or trees:

Stamens the same number as and opposite to mostly petaloid calyx (perianth)-segments; leaves often many-times divided; flowers often in heads or dense spikes or racemes *Proteaceae*

Stamens usually more numerous than the sepals or calyx-lobes, these rarely petaloid:

Leaves digitate; flowers unisexual *Euphorbiaceae*

Leaves pinnate or trifoliolate:

Flowers mostly polygamous; bark bitter; wood resinous; leaves pinnate *Simaroubaceae*

Flowers mostly polygamo-dioecious; bark not bitter; wood not resinous; leaves pinnate; stamens sometimes eccentric *Sapindaceae*

Flowers various; wood resinous; leaves pinnate or trifoliolate *Anacardiaceae*

Leaves simple:

Leaves modified into pitchers or tubes:

Flowers bisexual; stamens free; style entire, truncate at the apex *Sarraceniaceae*

Flowers dioecious; stamens united into a column *Nepenthaceae*

Leaves not modified as above:

Stamens circinately involute in bud; calyx-tube often rather long and then resembling a corolla; ovary 1-locular; ovule 1, basal, erect *Nyctaginaceae*

Stamens sometimes inflexed but not circinate in bud:

Stamens more or less connate into a central column:

Herbs or weak climbers; calyx (perianth) often rather long-tubular with an oblique (rarely actinomorphic) limb; ovules numerous; seeds with smooth endosperm; flowers bisexual *Aristolochiaceae*

Habit very various; calyx not oblique; flowers unisexual; ovules 1 or 2, pendulous; endosperm smooth *Euphorbiaceae*

Trees, shrubs or shrublets; ovules attached to the inner angle *Gyrostemonaceae*

Trees or shrubs; calyx mostly very small, not oblique; ovule erect; seeds with ruminate endosperm *Myristicaceae*

Stamens free or the filaments shortly connate only at the base:

Flowers arranged in a cyathium margined with glands; males consisting of 1 stamen, female single in the middle *Euphorbiaceae*

Flowers not as above; males with more than 1 stamen:

Stamens distinctly perigynous or flowers unisexual:
Twining plants; flowers in axillary spikes, racemes or panicles; ovule solitary, basal:
Anthers opening by slits lengthwise; embryo not spirally twisted; calyx spreading and reticulate-veiny in fruit; rootstock large and turnip-like *Agdestidaceae*
Anthers opening by terminal pores or pore-like slits; embryo spirally twisted *Basellaceae*
Herbs, not twining:
Calyx long and tubular, many-ribbed; stamens about 11, unequal; ovules 2 or more *Lythraceae*
Calyx-tube short or almost absent:
Ovary quite superior; leaves often thick and fleshy *Crassulaceae*
Ovary quite superior; leaves not thick and fleshy *Scrophulariaceae* (*Synthyris*)
Ovary semi-superior; leaves not fleshy *Aizoaceae* (*Ficoidaceae*)
Trees or shrubs; anthers opening by valves or longitudinal slits:
Small shrubs; stamens numerous, in bundles alternate with the calyx-lobes *Aizoaceae* (*Ficoidaceae*)
Mostly small shrubs with often rather long and usually petaloid calyx (perianth); leaves usually small and evergreen, sometimes much divided; stamens definite in number:
Calyx-lobes imbricate:
Fruit indehiscent *Thymelaeaceae*
Fruit a capsule *Aquilariaceae*
Calyx-lobes valvate:
Stamens the same number as and opposite the calyx-lobes *Proteaceae*
Stamens more than the primary lobes of the calyx; anthers inflexed in bud *Lythraceae*
Trees or large shrubs; leaves usually large:
Stamens numerous *Monimiaceae*
Stamens double the number of the calyx-lobes, in 2 or 4 rows *Lauraceae*
Stamens hypogynous or slightly perigynous if accompanied by a disk, or flowers unisexual:
Flowers in a cyathium margined by glands; male flowers reduced to 1 stamen *Euphorbiaceae*
Flowers not in a cyathium:
Trees or shrubs (to p. 79):
Leaves glandular; flowers in catkin-like spikes or racemes; berry often warted *Myricaceae*
Leaves not glandular; flowers not in catkin-like inflorences:
Leaves very small and ericoid; fruit drupaceous succulent; ovary 2-9-locular with 1 axile ovule in each loculus; stamens 2 or 3 *Empetraceae*

Leaves not as above and other characters not associated:
Ovules 6-8, pendulous *Peridiscaceae*
Ovules 2 or 1:
Flowers unisexual:
Ovule 1, pendulous; seeds with straight embryo *Trimeniaceae* and *Daphniphyllaceae*
Ovules 2, axile *Sapindaceae*
Ovule 1, erect *Achatocarpaceae*
Flowers bisexual; ovule 1, pendulous; embryo straight *Olacaceae*
Flowers bisexual; ovule 1, basal, erect *Barbeuiaceae*
Herbs (from p. 78):
Ovary of several subdistinct carpels; ovule 1 in each carpel, basal *Phytolaccaceae*
Ovary of united carpels:
Calyx (perianth) more or less scarious *Amaranthaceae*
Calyx (perianth) herbaceous:
Ovary 1-locular *Chenopodiaceae*
Ovary more than 1-locular *Molluginaceae*

'SYNCARPAE'. **'Axiles'.** Group 26. *Gynoecium composed of* 1 *carpel or of* 2 *or more united carpels with free or united styles, or if carpels free below, then the styles or stigmas united; ovules attached to the central axis or to the base or apex of the ovary; ovary superior, rarely partly immersed in the disk; petals absent; bisexual or male flowers (and often the female) with a distinct calyx; leaves opposite or verticillate.*

Leaves with stipules (to p. 81):
- **Stamens** more than twice the number of the sepals or calyx-lobes:
  - **Flowers** unisexual:
    - **Style** or styles terminal — *Euphorbiaceae*
    - **Style** basal; male sepals 2; embryo horse-shoe-shaped; lower leaves opposite, upper alternate — *Cynocrambaceae*
  - **Flowers** bisexual:
    - **Sepals** free from each other or nearly so; stamens hypogynous:
      - **Filaments** trifid at the apex, the two lateral lobes overlapping the anther; ovary 10-locular; stamens 10 — *Zygophyllaceae*
      - **Filaments** entire:
        - **Ovary** 5-locular; stigmas ligulate, thick — *Geraniaceae*
        - **Ovary** 1-5-locular; stigmas not ligulate:
          - **Herbs** or slightly woody at the base; leaves not compound — *Molluginaceae*
          - **Trees;** stipules large and membranous; leaves sometimes compound — *Cunoniaceae*
    - **Sepals** united into a tube; stamens inserted on the tube, more or less perigynous:
      - **Fruit** a capsule — *Aizoaceae* (*Ficoidaceae*)
      - **Fruit** an achene — *Rosaceae*
- **Stamens** up to twice as many as the sepals:
  - **Stamens** the same number as and alternate with the sepals:
    - **Trees** or shrubs; leaves not fleshy, often toothed — *Rhamnaceae*
    - **Herbs** or undershrubs with fleshy entire leaves — *Aizoaceae* (*Ficoidaceae*)
  - **Stamens** the same number as and opposite to the sepals or more numerous or fewer:
    - **Ovary** 1-locular with free central placentation; ovules usually several to numerous; stipules when present often dry and membranous — *Caryophyllaceae*
    - **Ovary** with axile, basal, or apical placentation, sometimes only 1 ovule:
      - **Flowers** arranged in an involucre (cyathium) margined with fleshy often more or less semilunar glands; male flowers reduced to 1 stamen, jointed about the middle; ovary often stalked, mostly 3-locular — *Euphorbiaceae*
      - **Flowers** not arranged as above; stamens usually more than 1; ovary mostly sessile:
        - **Flowers** unisexual (to p. 81):
          - **Ovary** 2- or more-locular; ovules 1 or 2, axile — *Euphorbiaceae*

Ovary 1-locular:

Ovule erect:

Filaments inflexed in bud; mostly herbaceous plants with fibrous stems and sometimes stinging hairs and leaves often with cystoliths; juice not milky *Urticaceae*

Filaments not inflexed in bud; mostly trees and shrubs, very rarely herbs; juice mostly milky; flowers often minute and arranged on or inside an enlarged 'receptacle' *Moraceae*

Ovule pendulous:

Filaments not inflexed in bud; erect or climbing herbs; flowers dioecious, males paniculate, female spicate *Cannabaceae*

Filaments inflexed in bud; large trees with milk-like juice; male flowers in dense cylindric spikes, females globose-capitate *Moraceae*

Flowers bisexual (from p. 80):

Sepals free or nearly completely so; stamens more or less hypogynous or perigynous:

Ovary 2-10-locular:

Ovules numerous in each loculus:

Calyx not greatly enlarged in fruit:

Stipules often lobed or laciniate, persistent; mostly herbs; flowers not capitate *Molluginaceae*

Stipules membranous, deciduous; trees; flowers capitate *Cunoniaceae*

Calyx greatly enlarged in fruit *Cunoniaceae*

Ovules 2 or few in each loculus; ovary 5-10-locular; stamens 5 or 10; flowers not capitate:

Filaments 3-fid at the apex; ovary 10-locular; ovules 2 or 3 in each loculus, ascending; leaves opposite, connate at the base, subterete, very juicy *Zygophyllaceae* (*Augea*)

Filaments entire; ovary 5-locular, beaked; ovules oendulous from the inner angle of the loculi; leaves opposite, entire or divided *Geraniaceae*

Ovary 1-locular:

Stipules forming a tube (ochrea) around the stem; fruit a nut, indehiscent *Polygonaceae*

Stipules not forming a tube around the stem; often scarious and bilobed or more-divided; leaves opposite *Illecebraceae*

Stipules minute; leaves verticillate *Molluginaceae*

Sepals connate into a tube; stamens perigynous; ovary 1-5-locular:

Flowers bisexual; leaves simple *Aizoaceae* (*Ficoidaceae*)

Flowers dioecious; leaves 1-3-foliolate *Rosaceae*

Leaves without stipules (from p. 80):

Ovary 1-locular, with free-central placentation on a central axis, not septate or only imperfectly so at the base; ovules usually numerous; mostly herbs *Caryophyllaceae*

Ovary with axile, basal, or apical placentation; ovule sometimes 1:

Ovary and fruit compressed contrary to the septum, 2-locular; trees or shrubs with perulate buds; leaves simple or compound:

Stamens more than 2; anther-loculi side by side *Aceraceae*
Stamens 2; anther-loculi back to back *Oleaceae*
Ovary if compressed then not contrary to the septum or septa:
Ovules 2 or more in each loculus or in a 1-locular ovary (to p. 83):
Flowers bisexual:
Shrubs or trees with woody branches, sometimes thorny:
Calyx more or less spreading or at any rate not very tubular:
Large disk often present *Celastraceae*
Disk absent:
Calyx-segments petaloid; ovary 4-locular; ovules 2 in each loculus, pendulous from the apex; South Africa *Geissolomataceae*
Calyx of free sepals, not petaloid; ovary 5- or 3-locular *Ledocarpaceae*
Calyx tubular, lobes valvate; disk absent or very inconspicuous:
Ovary many-locular, with numerous ovules; fruit a berry with numerous seeds *Sonneratiaceae*
Ovary 4-2-locular:
Calyx not or rarely somewhat petaloid, often with accessory lobes *Lythraceae*
Calyx shortly tubular, without accessory lobes; stamens 2 *Oleaceae*
Calyx more or less petaloid:
Stamens opposite the calyx (perianth)-lobes; flowers often in heads *Proteaceae*
Stamens alternate with the calyx-lobes; small shrubs of ericoid habit; fruit a loculicidal capsule; South Africa *Penaeaceae*
Herbs:
Sepals free or nearly so; stamens usually hypogynous:
Leaves connate at the base; *Illecebraceae*
Leaves not connate at the base:
Stamens opposite the calyx-lobes; embryo annular *Amaranthaceae*
Stamens alternate with the calyx-lobes; embryo straight *Primulaceae* (*Glaux*)
Sepals united into a tube; stamens perigynous:
Calyx-lobes imbricate, without accessory lobes *Aizoaceae* (*Ficoidaceae*)
Calyx-lobes valvate, often with accessory lobes; some aquatics *Lythraceae*
Flowers unisexual or polygamo-dioecious:
Leaves digitately compound *Euphorbiaceae*
Leaves simple:
Inflorescence a pendulous catkin; ovary 1-locular, inferior *Garryaceae*
Inflorescence not a catkin; lovary superior:
Fruit a drupe *Oleaceae*
Fruit a capsule:
Stamens 4, opposite the sepals *Buxaceae*

Stamens 4 or 5, alternate with the calyx-lobes *Crypteroniaceae*
Ovule single in each loculus of the ovary or in a 1-locular ovary:
Stamens circinately involute in bud; calyx usually long-tubular, often coloured and resembling a corolla; bracts sometimes petaloid; ovule basal, erect *Nyctaginaceae*
Stamens sometime inflexed but not circinate in bud; calyx often scarious but rarely petaloid; bracts often scarious:
Submerged aquatics with verticillate divided leaves; flowers monoecious *Ceratophyllaceae*
Not submerged aquatics; leaves very rarely verticillate:
Flowers not arranged in a cyathium; mostly bisexual:
Trees, shrubs, or undershrubs with woody or very tough stems; embryo straight, usually very small:
Stamens numerous:
Anthers opening from the base upwards by valves; style present:
Seeds with endosperm carpels separate in fruit; anthers 2-locular *Monimiaceae*
Seeds without endosperm; carpels not separate in fruit *Lauraceae*
Anthers opening by slits lengthwise; stigma sessile *Trimeniaceae*
Stamens 10 or fewer:
Indumentum lepidote; stamens 8; flowers dioecious *Elaeagnaceae*
Indumentum not lepidote or absent:
Calyx deeply lobed, the lobes of the female much enlarged in fruit and venose; trees; seeds with endosperm; stamens 6-9 *Barbeyaceae*
Calyx usually tubular, not 2-lipped; flowers often in heads; stamens sometimes reduced to 1 *Thymelaeaceae*
Calyx campanulate, membranous, 2-lipped; flowers densely spicate, dioecious; fruit a fleshy syncarp; stigmas sessile *Batidaceae*
Herbs, sometimes rather woody at the base; embryo more or less curved:
Sepals united into a tube; stamens perigynous *Aizoaceae* (*Ficoidaceae*)
Sepals free or nearly so; stamens mostly hypogynous:
Calyx hyaline only on the margin; stamens hypogynous; styles free or nearly so; flowers sometimes unisexual *Molluginaceae*
Calyx often hyaline all over; stamens often slightly perigynous and connate at the base *Amaranthaceae*
Calyx herbaceous (Greenish); stamens hypogynous or slightly perigynous, mostly free *Chenopodiaceae*
Flowers arranged in a cyathium margined with glands (rather resembling a single flower with numerous stamens and one ovary; *Euphorbiaceae*

'SYNCARPAE'. **'Axiles'.** Group 27. *Gynoecium composed of* 1 *carpel or of* 2 *or more united carpels with free or united styles, or if carpels free below, then the styles or stigmas united; ovules attached to the central axis or to the base or apex of the ovary; ovary superior, rarely partly immersed in the disk; petals absent; bisexual or male flowers* (*and also the female*) *without a distinct calyx* (*perianth*).

Fleshy nevergreen herbs parasitic on roots of other plants and devoid of green colouring matter; ovule solitary in a 1-locular ovary: *Balanophoraceae*
Moss-like or hepatic-like aquatic herbs with minute flowers; ovules 2 or more in a 1-locular ovary *Podostemaceae*
Neither parasites nor aquatic plants with above habits:
Leaves all radical, palminerved, 3-partite; flowers in a slender spike; carpel 1; ovule 1, erect from the base of the loculus; herbs with creeping rhizomes *Podophyllaceae*
Leaves not all radical and above characters not associated:
Trees or shrubs with whorled scale-like leaves and jointed branches; stamen 1, filament lengthened during flowering *Casuarinaceae*
Trees, shrubs, shrublets or herbs with normally developed leaves;
Leaves stipulate; stipules sometimes adnate to the petiole (to p. 85):
Ovary 1-locular:
Herbs, shrubs or trees; leaves usually alternate; flowers in dense spikes or catkins; petiole not dilated and not enclosing the young bud:
Leaves not aromatic and not glandular; vascular bundles scattered as in Monocotyledons; seeds with copious endosperm *Piperaceae*
Leaves aromatic, often very glandular; vascular bundles in rings; seeds without endosperm *Myricaceae*
Trees or shrubs with alternate leaves; flowers minute on a common open receptacle with the fruits becoming immersed in it *Moraceae*
Herbs, shrubs or trees with opposite leaves *Chloranthaceae*
Large trees; leaves alternate, palmately nerved and lobed, the dilated petiole enclosing the young bud; anthers with a peltate connective at the apex; flowers monoecious, densely arranged in heads *Platanaceae*
Ovary 2- or more-locular; leaves alternate, rarely opposite:
Ovules numerous:
Stipules not spiny; large trees; leaves not plicately nerved *Hamamelidaceae*
Stipules subulate-spiny when old; leaves opposite, plicately nerved and folded in bud; low shrubs in South and south-tropical Africa *Myrothamnaceae*
Ovules 1 or 2 in each locules of the ovary; stipules rarely spiny, often caducous:
Flowers very variously arranged, sometimes in an involucre (cyathium) margined with fleshy glands; seeds usually with copious endosperm; ovary mostly 2- or 3-locular or sometime more-locular *Euphorbiaceae*

Male flowers in catkins or slender spikes; endosperm absent from the seeds; ovary 2-6-locular:

Fruits subtended by a cupular involucre or enclosed by the latter *Fagaceae*

Fruits cone-like, with imbricate scales *Betulaceae*

Fruits separate, not arranged as above; leaves trifoliolate *Picrodendraceae*

Leaves without stipules (from p. 84):

Trees or shrubs with hard wood; vascular bundles in rings (not scattered as in Monocotyledons):

Flowers in a cyathium; male flowers reduced to 1 stamen; stems with milky juice *Euphorbiaceae*

Flowers not in a cyathium:

Leaves not aromatic; flowers dioecious:

Male flowers in catkins on the young branchlets or below the leaves, the female single in an involucre of bracts *Balanopsidaceae*

Male and female flowers solitary, precocious; fruit samaroid, 1-seeded *Eucommiaceae*

Male flowers in short spikes, female often solitary; fruit not samaroid *Buxaceae*

Leaves aromatic, often glandular; flowers in dense spikes; mostly swamp plants or on sand-dunes; leaves simple *Myricaceae*

Leaves not aromatic, simple or pinnate; not swamp plants; ovary 2-locular; stamens 2 or 4 *Oleaceae*

Leaves not aromatic, fleshy; flowers dioecious; ovary 4-locular; stamens 4; leaves simple; maritime plants *Batidaceae*

Herbs or shrubs, erect or scandent, sometimes slightly woody at the base:

Ovary 1-locular:

No stinging hairs; stems not fibrous; epidermal cells without cystoliths; vasular bundles not scattered *Chenopodiaceae*

No stinging hairs; stems not fibrous; flowers minute, usually densely spicate; epidermal cells often with cystoliths; vascular bundles more or less scattered as in Monocotyledons *Piperaceae*

Stinging hairs often present; stems often fibrous; epidermal cells mostly with prominent cystoliths; vascular bundles not scattered *Urticaceae*

Ovary 2- or 3-locular; flowers much reduced, often in a cyathium with glands on the margin *Euphorbiaceae*

Ovary 4-locular; annual herbs with minute unisexual flowers; styles 2; stamen 1 *Callitrichaceae*

'SYNCARPAE'. **'Axiles'.** Group 28. *Gynoecium composed of* 1 *carpel or of* 2 *or more united carpels with free or united styles, or if carpels free below, then the styles or stigmas united; ovules attached to the central axis or to the base or apex of the ovary; ovary more or less inferior to semi-inferior; petals present, free from each other; leaves alternate, spirally arranged or all radical.*

Flowers bisexual (to p. 89):
Stamens numerous (to foot of this page):
Aquatic herbs with floating leaves; flowers usually large and showy; ovules numerous *Nymphaeaceae*
Not aquatic; habit various:
Herbs:
Leaves without stipules; sepals more than 2:
Styles more or less free; low herbs often with rosettes of leaves *Saxifragaceae*
Styles more or less united; erect or climbing, mostly roughly hispid-pilose herbs *Loasaceae*
Leaves stipulate, stipules often laciniate and thread-like; seeds with more or less copious endosperm; sepals 2 *Portulacaceae*
Leaves stipulate, stipules not laciniate; seeds without endosperm; sepals more than 2 *Rosaceae*
Trees, shrubs or shrublets:
Leaves gland-dotted or pellucid-punctate; stamens mostly very numerous:
Fruits not winged; stamens regularly arranged in similar whorls or in 5 bundles opposite the petals *Myrtaceae*
Fruits broadly winged; stamens numerous, in several series, filaments slightly connate at the base *Barringtoniaceae* (*Combretodendron*)
Leaves neither gland-dotted nor punctate:
Leaves with stipules:
Styles more or less divided or styles separate:
Stipules not intrapetiolar; leaves not digitate, sometimes pinnate *Rosaceae*
Stipules intrapetiolar and connate; leaves digitate *Araliaceae*
Style simple; plants usually maritime; leaves simple *Rhizophoraceae*
Leaves without stipules:
Stamens free or slightly connate at the base, not in 1 or 2 bundles:
Petals free; endosperm of the seeds ruminate *Alangiaceae*
Petals free; endosperm absent or subruminate *Barringtoniaceae*
Petals more or less united; endosperm not ruminate *Styracaceae*
Stamens very numerous and more or less mona- or diadelphous, if the latter then one bundle ligulate and either bearing anthers or barren and more or less arched over the gynoecium; seeds without endosperm *Lecythidaceae*
Stamens definite in relation to the sepals and petals, the same number as or about twice as many, rarely fewer:

Stamens the same number as and opposite the petals:
- **Leaves** stipulate or not; not parasites; flowers often crowded into heads; ovary 3-locular:
  - **Fruit** not enclosed by the calyx; ovules erect from the base of the ovary; ovary immersed in the disk *Rhamnaceae*
  - **Fruit** enclosed by the greatly enlarged calyx; ovules pendulous from the apex of the ovary *Erythropalaceae*
- **Leaves** without stipules; plants often parasitic or semi-parastitic; ovary 1-locular:
  - **Leaves** mostly opposite, sometimes reduced to scales; calyx often much reduced; ovules usually not clearly differentiated from the placenta *Loranthaceae*
  - **Leaves** alternate; calyx distinct, lobes imbricate or open in bud; ovules clearly differentiated from the placenta *Olacaceae*

Stamens the same number as and alternate with the petals or more numerous or fewer:
- **Anthers** opening by apical pores:
  - **Leaves** often with very prominent longitudinal nerves parallel with the midrib and margin, not gland-dotted; anthers often unequal, with the connective usually produced at the base *Melastomataceae*
  - **Leaves** and nerves not as above; gland-dotted; anthers equal *Myrtaceae*
- **Anthers** opening by slits lengthwise; nervation of the leaves usually pinnate:
  - **Leaves** stipulate:
    - **Herbs:**
      - **Stipules** setaceous or lacerate *Saxifragaceae*
      - **Stipules** not as above:
        - **Leaves** simple: sepals 2; fruit a capsule *Portulacaceae*
        - **Leaves** simple; calyx-lobes 5, with an epicalyx of bracteoles *Rosaceae*
        - **Leaves** usually compound or much dissected, sometimes peltate; sepals obsolete or more than 2; fruit of 2 indehiscent mericarps *Apiaceae* (*Umbelliferae*)
    - **Trees** or shrubs or woody climbers:
      - **Flowers** actinomorphic ('regular'); petals not gibbous at the base; stamens not unilaterial:
        - **Leaves** simple; stipules paired; flowers often capitate and sometimes precocious; fruit woody, mostly a capsule; stamens as many as the petals, anthers mostly opening by valves *Hamamelidaceae*
        - **Leaves** compound, rarely simple; stamens as many as the petals; fruit a berry or drupaceous; flowers often umbellate *Araliaceae*
        - **Leaves** compound or simple; stamens more than the petals; fruit not woody:
          - **Filaments** not toothed at the top *Rosaceae*
          - **Filaments** broad and toothed at the top *Pterostemonaceae*
      - **Flowers** zygomorphic ('irregular'); petals often gibbous at the base; stamens unilateral *Trigoniaceae*
  - **Leaves** without stipules:

Flowers arranged in heads surrounded by a brightly coloured involucre simulating a single flower; petals very unequal, fewer than the stamens; styles subulate, elongated *Hamamelidaceae*
Flowers not arranged as above and other characters not associated if in heads:
Herbs:
Flowers umbellate; ovules solitary, pendulous:
Ovary 2-locular; carpels separating in fruit into 2 indehiscent mericarps; filaments inflexed in bud *Apiaceae* (*Umbelliferae*)
Ovary 3- or 4-locular; carpels not separating in fruit *Araliaceae*
Flowers not umbellate:
Ovules more than 1 in each loculus or in a 1-locular ovary:
Mostly scapigerous herbs; ovary 1-3-locular:
Fruits capsular; flowers rarely solitary:
Anthers all glabrous *Saxifragaceae*
Anthers partly setose, partly glabrous *Campanulaceae*
Fruits indehiscent; flowers solitary ovules on placentas near the top of the central axis *Donatiaceae*
Leafy-stemmed herbs:
Ovary mostly 4-locular of united carpels; petals contorted *Onagraceae*
Ovary of partly free carpels; calyx 3-lobed *Aristolochiaceae*
Ovule solitary in each loculus of the ovary:
Style 1, with a capitate or shortly lobed stigma; flowers mostly fairly large and conspicuous:
Plants neither rough-setaceous nor scabrid; ovules 1 or more in the whole ovary *Onagraceae*
Plants usually roughly setose or scabrid; ovule solitary from the apex of a 1-locular ovary *Loasaceae*
Styles more than 1:
Styles 5–10; carpels 5–10, radiating; stigmas capitate *Rosaceae*
Styles up to 4; ovary up to 4-locular; carpels not radiating; stigmas not capitate *Haloragidaceae*
Styles 2; ovary about $\frac{3}{4}$ superior, 2-locular; ovules erect *Eremosynaceae*
Shrubs, shrublets, trees or climbers:
Flowers umbellate; fruit a berry or drupe *Araliaceae*
Flowers not umbellate:
Leaves 3-foliolate; petals 5; stamens 5, alternating with glands; anthers opening by valves *Hernandiaceae*
Leaves simple; anthers opening by slits:
Petals contorted; stamens mostly 4 or 8 *Onagraceae*
Petals valvate or imbricate:
Alternate leaves not larger and smaller:
Filaments not toothed at the top:
Petals valvate, loriform *Alangiaceae*
Petals more or less imbricate, if valvate not loriform:
Trees or shrubs:

Fruit a capsule or berry *Escalloniaceae*
Fruit a drupe *Cornaceae*
Climbers; fruit indehiscent, 1-seeded *Loasaceae*
Filaments broad and toothed at the top *Pterostemonaceae*
Alternate leaves larger and smaller (anisophyllous); petals convolute or inflexed *Rhizophoraceae*

Flowers unisexual or polygamous (from p. 86):
Flowers not in heads or umbels, sometimes paniculate or racemose:
Leaves stipulate, often very unequal sided at the base; stipules paired; no tendrils; stamens mostly numerous, straight; flowers often somewhat zygomorphic *Begoniaceae*
Leaves without stipules; tendrils often present; stamens definite in number or rarely many; anthers often conduplicate or twisted, mostly 3; flowers actinomorphic, conspicuous *Cucurbitaceae*
Leaves without stipules or if these present then adnate to the petiole, sometimes anisophyllous and then the smaller leaf appearing like a stipule; no tendrils; anthers straight:
Trees, shrubs or shrublets:
Leaves gland-dotted or pellucid-punctate *Myrtaceae*
Leaves not gland-dotted:
Trees or shrubs with dioecious flowers and palmately or pinnately nerved leaves, these rarely spiny-dentate *Cornaceae*
Shrubs with polygamous flowers and pinnately nerved leaves *Escalloniaceae*
Mostly herbs, sometimes with very large leaves, often aquatic or near water; flowers mostly very minute *Haloragidaceae*
Flowers arranged in heads, umbels, or corymbs:
Fruits of dry indehiscent mericarps; ovary 2-locular; styles 2; filaments inflexed in bud; herbs with usually much dissected leaves *Apiaceae* (*Umbelliferae*)
Fruits and other characters not combined as above:
Anthers opening by a single lateral valve; petals linear-spathulate; ovary 2-locular; ovules solitary or numerous; leaves stipulate *Hamamelidaceae*
Anthers opening by slits; petals usually not linear-spathulate:
Leaves usually stipulate; flowers mostly umbellate, rarely epiphyllous *Araliaceae*
Leaves without stipules; flowers capitate or corymbose:
Petals imbricate *Nyssaceae*
Petals valvate; flower-heads often with an involucre of petaloid bracts *Cornaceae*
Leaves without stipules; male flowers corymbose, female solitary *Onagraceae*

'SYNCARPAE'. **'Axiles'.** Group 29. *Gynoecium composed of* 1 *carpel or of* 2 *or more united carpels with free or united styles, or if carpels free below, then the styles or stigmas united; ovules attached to the central axis or to the base or apex of the ovary; ovary more or less inferior; petals present, free from each other; leaves opposite or verticillate, rarely reduced to scales.*

Leaves simple (sometimes very deeply divided) (to p. 92):
Leaves stipulate:
Stamens the same number as and opposite the petals *Rhamnaceae*
Stamens alternate with the petals or more numerous:
Ovary composed of 2 carpels more or less free at the apex; flowers mostly capitate; ovules pendulous; trees or shrubs; anthers often opening by valves *Hamamelidaceae*
Ovary mostly 2-locular with 2 simple styles; flowers sometimes but rarely capitate; anthers opening by slits lengthwise *Cunoniaceae*
Ovary composed of 2-6 carpels, 2-6-locular or 1-locular by suppression of the septa; flowers rarely congested:
Ovules pendulous; stipules interpetiolar:
Flowers not in umbels; disk perigynous; trees or shrubs, sometimes maritime *Rhizophoraceae*
Flowers in umbels; disk epigynous *Araliaceae*
Ovules spreading from the axis; stipules intrapetiolar *Dialypetalanthaceae*
Ovary various; flowers rarely in heads; ovules ascending or attached to the central axis; trees shrubs or herbs *Rosaceae*
Leaves without stipules:
Trees, shrubs or climbers, sometimes parasitic, rarely prostrate (to p. 91):
Stamens numerous:
Ovary-loculi not superposed:
Leaves gland-dotted, often ericoid; style simple with a small capitate stigma or very rarely 3- or 4-lobed; filaments free or united into bundles sometimes opposite the petals *Myrtaceae*
Leaves with pellucid lines or often with stellate hairs; style 5-10-lobed or styles 2-5 and more or less free; ovary 2-10-locular *Philadelphaceae*
Leaves not gland-dotted, with prominent longitudinally parallel nerves; stamens jointed or kneeded with the connective often produced at the base; anthers opening by a terminal pore *Melastomataceae*
Ovary-loculi superposed:
Leaves not gland-dotted; style simple *Punicaceae*
Leaves gland-dotted *Myrtaceae*
Stamens as many to twice as many as the petals:
Stamens the same number as and opposite the perianth-segments (petals) or lobes (to p. 91):
Mostly parasitic or semi-parasitic shrubs, or trees; calyx usually much reduced *Loranthaceae*

Habit not as above; calyx distinct; leaves glandular-punctate *Myrtaceae*

Stamens the same number as and alternate with the petals or perianth-segments, or more numerous (from p. 90):

Anthers opening by terminal pores:

Filaments often jointed; leaves mostly with prominent longitudinal parallel nerves *Melastomataceae*

Filaments not jointed; leaves without longitudinal parallel nerves, often ericoid and gland-dotted *Myrtaceae*

Anthers opening by slits lengthwise; calyx mostly valvate:

Ovule solitary in each loculus; fruit mostly drupaceous; endosperm copious; petals without alternate scales:

Stipules present; flowers mostly in simple umbels, rarely racemose or capitate *Araliaceae*

Stipules absent:

Leaves mostly entire; disk cushion-shaped, central in the male flowers, epigynous in the female *Cornaceae*

Leaves mostly with gland-tipped teeth; disk lobed, lobes alternate with the stamens *Escalloniaceae*

Ovules 2 or more in each loculus:

Stamens erect in bud:

Leaves gland-dotted *Myrtaceae*

Leaves not gland-dotted:

Stamens 8 or more; fruit a loculicidal capsule, rarely a berry:

Filaments often toothed near the apex; flowers in racemes, cymes, heads or panicles; indumentum mostly of stellate hairs *Philadelphaceae*

Filaments not toothed; flowers cymose or corymbose; indumentum absent or of simple hairs *Hydrangeaceae*

Stamens 5 or 4; fruit a drupe or berry:

Leaves distant and not small or heath-like; style 1; flowers in short cymes *Oliniaceae*

Leaves mostly crowded and heath-like; styles 2 or 3, partly cohering; flowers in dense spikes or heads *Bruniaceae*

Stamens inflexed in bud; fruits often winged *Combretaceae*

Herbs, sometimes slightly woody but then often with fleshy leaves (from p. 90):

Anthers opening by terminal pores; leaves mostly with prominent longitudinally parallel nerves; connective of the anthers usually produced at the base and jointed to the filament *Melastomataceae*

Anthers opening by slits lengthwise:

Flowers in umbels or heads; ovary 2-locular; carpels separating in fruit and suspended by the divided thread-like central axis (carpophore) *Apiaceae* (*Umbelliferae*)

Flowers in heads surrounded by bracts; ovary 2- or 3-locular; carpels not as above *Cornaceae*

Flowers not in umbels or heads; carpels not separating in fruit:

Placentas pendulous from the apex of the 1-locular ovary; ovules numerous *Vahliaceae*

Placenta pendulous from the top of the ovary-loculi; ovules 1-4 *Haloragidaceae*

Placentas axile with numerous ovules, or ovules very few and pendulous from the apex of the usually 1-2-4-locular ovary; calyx-lobes valvate:

Cotyledons equal; plants rarely aquatic; fruits not spiny; stamens and petals sometimes 2 each *Onagraceae*

Cotyledons very unequal; floating herbs with spiny fruits and dimorphous leaves; petals and stamens 4 each *Trapaceae*

Placentas at the bottom of the ovary or ovary-loculi; petals numerous; calyx-lobes imbricate *Aizoaceae* (*Ficoidaceae*)

Leaves compound, much divided or unifoliolate; flowers mostly umbellate or capitate, rarely racemose; calyx small, entire or toothed; petals usually 5, valvate or slightly imbricate; disk on top of the ovary, often confluent with the style or styles; ovule solitary in each loculus, pendulous; seeds with copious endosperm and small embryo (from p. 90):

Trees, shrubs or climbers; petals usually valvate; ovary 1-many-locular; fruit usually a berry or drupe *Araliaceae*

Herbs often with hollow stems between the nodes; petals imbricate, rarely valvate; ovary 2-locular; styles 2; fruit of 2 dry indehiscent mericarps *Apiaceae* (*Umbelliferae*)

'SYNCARPAE'. **'Axiles'.** Group 30. *Gynoecium composed of* 1 *carpel or of* 2 *or more carpels united with free or united styles, or if carpels free below then the styles or stigmas united; ovules attached to the central axis or to the base or apex of the ovary; ovary more or less inferior; petals present, more or less united into a tube, cup or cap; leaves alternate or all radical.*

Stamens 3 or more (to p. 95):
Anthers free from one another or very slightly connate only at the base (to p. 94):
Stamens the same number as and opposite the corolla-lobes:
Herbs; leaves not gland-dotted; staminodes present between the stamens *Primulaceae* (*Samolus*)
Trees and shrubs:
Leaves often gland-dotted; plants not parasitic *Myrsinaceae*
Leaves not gland-dotted; plants often parasitic *Loranthaceae*
Stamens alternate with the corolla-lobes or more numerous or fewer:
Corolla actinomorphic ('regular') (to p. 94):
Herbaceous plants, often climbers, some with milk-like juice:
Flowers unisexual; stems usually climbing by means of tendrils; anthers often sinuous or twisted *Cucurbitaceae*
Flowers bisexual or very rarely unisexual; no tendrils; anthers usually straight:
Leaves with stipules; sepals 2 *Portulacaceae*
Leaves without stipules; sepals or calyx-lobes 2 or more:
Anthers 2-locular:
Filaments free or nearly so; ovary usually 2- or more-locular *Campanulaceae*
Filaments more or less connate; anthers more or less connate around the style; ovary 1-locular, style slender, undivided *Calyceraceae*
Anthers 1-locular due to the splitting of the filaments *Adoxaceae*
Woody plants, trees, shrubs or shrublets, rarely subherbaceous but juice not milk-like:
Leaves gland-dotted; stamens mostly numerous *Myrtaceae*
Leaves not gland-dotted:
Stamens free from the corolla or very slightly adherent to it only at the base (to p. 94):
Corolla-lobes valvate; anthers not produced into tubes (to p. 94):
Stipules present, often adnate to the petiole; stamens not very numerous *Araliaceae*
Stipules absent; corolla multiradiate with undulate margin (like an umbrella); stamens very numerous *Asteranthaceae*
Stipules absent; corolla campanulate; stamens 5 *Prionotaceae**

Corolla-lobes imbricate; calyx not wing-like in fruit:
Anthers opening by terminal pores *Vacciniaceae*
Anthers opening by slits lengthwise;
Stipules absent; leaves simple; flowers not in umbels, usually solitary *Theaceae*
Stipules present; flowers mostly in simple umbels, rarely racemose or capitate *Araliaceae*
Corolla-lobes contorted; small stipules present; branches hooked; calyx-lobes becoming enlarged and wing-like in fruit *Ancistrocladaceae*
Stamens inserted on the corolla (from p. 39):
Flowers in a terminal leafy capitulum; ovules 2, pendulous, collateral *Bruniaceae*
Flowers not in a capitulum:
Ovary 3-5-locular:
Stamens numerous: filaments free or partly connate; indumentum neither stellate nor lepidote *Symplocaceae*
Stamens mostly 5 or 10, filaments not united at the base; indumentum often stellate or lepidote *Styracaceae*
Stamens 5, filaments slightly adante at the base to the corolla; indumentum not as above *Prionotaceae**
Stamens 8, filaments shortly united; anthers apiculate *Lissocarpaceae*
Ovary 1-2-locular:
Ovule solitary, pendulous; flowers in cymes *Alangiaceae*
Ovules numerous: flowers fasciculate or solitary *Caprifoliaceae*
Corolla zygomorphic ('irregular') (from p. 93):
Ovule solitary, pendulous; stigma not indusiate *Valerianaceae*
Ovule 1 or more, erect or ascending; stigma indusiate *Goodeniaceae*
Ovules numerous, spreading on axile placentas; stigma not indusiate *Lobeliaceae*
Anthers more or less united into a ring or tube around the style (from p. 93):
Anthers straight, not twisted or flexuous:
Flowers not in heads surrounded by a common involucre of bracts:
Flowers bisexual, mostly zygomorphic; stigma indusiate *Goodeniaceae*
Flowers bisexual, zygomorphic; stigma not indusiate; no tendrils; habit varying from giant tree-like herbs to tiny herbs *Lobeliaceae*
Flowers unisexual, actinomophic; tendrils usually present; often climbers *Cucurbitaceae*
Flowers in heads surrounded by a common involucre of bracts; leaves without stipules:
Ovule erect from the base of the ovary-loculus; filaments usually free from each other; calyx often much modified into a pappus, or reduced or scale-like *Asteraceae* (*Compositae*)
Ovule pendulous from the top of the ovary-loculus; filaments more or less connate; calyx not modified as above *Calyceraceae*
Anthers flexuous or conduplicate; plants often climbing by means of tendrils *Cucurbitaceae*

Stamens 1 or 2 (from p. 93):
Flowers bisexual or unisexual by abortion; no tendrils; corolla often more or less zygomorphic:
Anthers with a narrow connective or the loculi divergent; filaments free from each other *Gesneriaceae*
Anthers with a very broad connective; filaments free from each other *Columelliaceae*
Anthers with a narrow connective; filaments connate *Stylidiaceae*
Flowers unisexual; mostly climbers with tendrils; corolla actinomorphic *Cucurbitaceae*

'SYNCARPAE'. **'Axiles'.** Group 31. *Gynoecium composed of* 1 *carpel or of* 2 *or more united carpels with free or united styles, or if carpels free below then the styles or stigmas united; ovules attached to the central axis or to the base or apex of the ovary; ovary more or less inferior; petals present, more or less united into a tube, cup or cap; leaves opposite or verticillate, never all radical.*

Leaves with stipules, these mostly inter- or intra-petiolar; anthers free from each other:
 Leaves simple, entire; corolla actinomorphic (very rarely slightly zygomorphic) *Rubiaceae*
 Leaves simple or pinnate, often toothed; corolla actinomorphic or zygomorphic; stipules very small or rarely large and foliaceous *Caprifoliaceae*
Leaves without stipules:
 Leaves with longitudinally parallel main nerves; stamens often double the number of the corolla-lobes, filaments mostly jointed, the connective produced at the base into an appendage *Melastomataceae*
 Leaves and stamens not in combination as described above:
  Anthers free from each other; ovules mostly pendulous, sometimes not distinct (to p. 97):
   Leaves gland-dotted; stamens mostly numerous; trees and shrubs usually with evergreen leaves *Myrtaceae*
   Leaves not gland-dotted; mostly herbaceous plants, rarely shrubs; stamens definite in number:
    Stigma indusiate; ovary 1- or 2-locular; flowers not capitate; corolla-lobes valvate *Goodeniaceae*
    Stigma not indusiate, often lobed:
     Herbaceous or woody only at the base:
      Flowers actinomorphic ('regular'); all loculi of the ovary fertile:
       Anthers 2-locular:
        Stamens usually 5 (more than 2) *Campanulaceae*
        Stamens 2 *Caprifoliaceae*
       Anthers 1-locular; stamens appearing to be double the number of the corolla-lobes due to division of the filaments *Adoxaceae*
      Flowers zygomorphic ('irregular'); fruits indehiscent:
       Ovary with 1 perfect 1-ovulate loculus and often 2 empty loculi; calyx-lobes sometimes becoming plumose *Valerianaceae*
       Ovary 1-locular, without additional empty loculi: flowers often capitate with a single or double involucel below the calyx *Dipsacaceae*
     Trees, shrubs, or woody climbers, or parasites on woody plants; corolla sometimes saccate or spurred at the base:
      Stamens double the number of the corolla-lobes; ovary 6-locular; a large tree *Lythraceae* (*Axinandra*)
      Stamens the same number as the corolla-lobes, rarely fewer:
       Stamens alternate with the usually imbricate corolla-lobes, sometimes

only 2 or 4; not parasitic, but sometimes trailing or climbing *Caprifoliaceae*

Stamens opposite the valvate corolla-lobes; plants often parasitic on branches of other woody plants *Loranthaceae*

Anthers mostly connate or connivent or in pairs around the style (from p. 96):

Ovules numerous; flowers usually not in heads:

Corolla actinomorphic ('regular'); stamens the same number as the corolla-lobes *Campanulaceae*

Corolla zygomorphic ('irregular'); stamens usually fewer (4 or 2) than the corolla-lobes *Gesneriaceae*

Ovule solitary; flowers mostly in heads surrounded by an involucre of bracts:

Ovule erect; calyx usually modified into a pappus of smooth, barbellate, or plumose bristles *Asteraceae* (*Compositae*)

Ovule pendulous; calyx not modified into a pappus *Calyceraceae*

'SYNCARPAE'. **'Axiles'.** Group 32. *Gynoecium composed of* 1 *carpel or of* 2 *or more united carpels with free or united styles, of if carpels free below then the styles or stigmas united; ovules attached to the central axis or to the base or apex of the ovary; ovary more or less inferior; petals absent.*

Plants not parasitic, or if so then more or less woody and often with normally developed green leaves (with chlorophyll) (to p. 100):
Leaves stipulate:
Flowers bisexual, often solitary or racemose:
Stamens the same number as and alternate with the sepals *Rhamnaceae*
Stamens the same number as and opposite to or more numerous than the sepals:
Leaves mostly alternate:
Leaves simple or compound; seeds without endosperm; no mangrove plants *Rosaceae*
Leaves never compound; trees and shrubs often with stellate indumentum; seeds with thin endosperm; no mangroves *Hamamelidaceae*
Leaves mostly opposite, rarely verticillate, trifoliolate or pinnate, rarely simple; seeds with copious endosperm; no mangroves *Cunoniaceae*
Leaves mostly opposite, mainly mangrove trees of the tropics; stamens often in pairs opposite the calyx-lobes *Rhizophoraceae*
Flowers unisexual; stipules often paired, sometimes very small:
Stamens and ovules numerous; fruits often winged; outer pair of calyx-lobes valvate; herbs, often with obliquely shaped leaves *Begoniaceae*
Stamens and ovules few, the latter solitary or paired, rarely numerous:
Trees or shrubs; stipules not adnate to the petiole:
Male flowers with a calyx (perianth):
Ovule 1 in the ovary; stamens often inflexed in bud *Moraceae*
Ovules 2 in each loculus of the ovary; stamens not inflexed in bud *Fagaceae*
Male flowers without a calyx (perianth):
Leaves alternate; stamens several:
Flowers mostly capitate; fruit capsular; seeds with thin fleshy endosperm *Hamamelidaceae*
Male flowers in catkins; fruit a nut enclosed in the foliaceous accrescent involucre; seeds without endosperm *Corylaceae*
Leaves opposite; petioles more or less connate at the base; stamens 1–3 *Chloranthaceae*
Herbs; stipules adnate to the petiole *Haloragidaceae*
Leaves without stipules, but sometimes when opposite the petioles are connate and sheathing at the base:
Flowers usually in catkins or slender spikes, or rarely the males in panicles, unisexual (to p. 99):
Leaves imparipinnate, alternate or rarely opposite; ovule 1, erect; fruit often adnate to the enlarged bracts or bracteoles *Juglandaceae*

Leaves simple:
Leaves alternate; ovules 1 or 2, descending *Fagaceae*
Leaves opposite; ovules pendulous:
Ovules 2 collateral; leaves with petioles connate at the base; flowers dioecious, arranged in silky catkin-like racemes *Garryaceae*
Ovule 1 in each loculus, pendulous; petioles more or less connate at the base; flowers spicate, paniculate or capitate *Chloranthaceae*
Flowers not in catkins, mostly bisexual (from p. 98):
Leaves opposite and gland-dotted *Myrtaceae*
Leaves alternate or if opposite not gland-dotted:
Flowers in simple or compound umbels; herbs with usually hollow stems between the nodes; leaves often sheathing at the base; ovary 2-locular, separating in fruit into 2 mericarps with resinous lines *Apiaceae* (*Umbelliferae*)
Flowers not in umbels or rarely so, sometimes in heads; fruit not as above:
Ovules numerous on axile placentas; calyx-lobes valvate or imbricate:
Stamens numerous; seeds usually without endosperm:
Trees or shrubs with alternate leaves *Barringtoniaceae**
Trees or shrubs with opposite leaves *Sonneratiaceae*
Stamens definite in number in relation to the sepals:
Sepals or calyx-lobes 10, imbricate; stamens 10, opposite the calyx-lobes; slender herbs; seeds with endosperm and minute embryo; ovary 2-locular *Saxifragaceae* (*Zahlbruchnera*)
Sepals or calyx-lobes valvate, 4–8:
Seeds without endosperm and large embryo; leaves opposite or alternate *Onagraceae*
Seeds with copious endosperm and small embryo; calyx actinomorphic or unilateral, often coloured; leaves alternate *Aristolochiaceae*
Ovules solitary or few, usually inserted at the top or base of the ovary, rarely axile or on a basal placenta:
Ovules pendulous from the apex of the ovary or at the apex of a basal placenta, usually more than 1 (to p. 100):
Trees, shrubs or woody climbers; sepals usually valvate (to p. 100):
Anthers opening by valves:
Ovary 1-locular:
Leaves simple; calyx-lobes imbricate *Lauraceae*
Leaves simple or compound, sometimes peltate; calyx-lobes valvate *Hernandiaceae*
Ovary 2- or 3-locular *Gomortegaceae*
Anthers opening lengthwise by slits:
Stamens erect in bud (to p. 100):
Stamens often double the number of the sepals *Nyssaceae*
Stamens the same number as and opposite the sepals or calyx-lobes:
Placenta not reaching to the top of the 1-locular ovary; indumentum not stellate *Santalaceae*
Placenta reaching to the top of the ovary and danate to it; indumentum stellate; flowers in axillary racemes or in panicles *Octoknemaceae*

Stamens with the filaments inflexed in bud, often double the number of the sepals (from p. 99):
Ovules inserted at the apex of the 1-locular ovary; flowers sometimes capitate; fruits often winged; calyx-lobes valvate *Combretaceae*
Ovules inserted on a free basal placenta of the at length 1-locular ovary *Grubbiaceae*
Herbaceous and often fleshy; sepals mostly imbricate *Aizoaceae* (*Ficoidaceae*)
Ovule 1, erect or axile (from p. 99):
Trees and shrubs, mostly maritime, with often lepidote leaves; flowers not capitate *Elaeagnaceae*
Not maritime; leaves not lepidote; flowers capitate or densely crowded; ovule well developed at flowering time *Rhamnaceae*
Twining herb with twisted petioles and large turnip-like rootstock; flowers paniculate; calyx- wing-like in fruit, venose-reticulate *Agdestidaceae*
Parasitic nevergreen herbs destitute of chlorophyll, the leaves reduced to scales or absent; ovules nude or with a single integument (from p. 98):
Ovules very numerous; flowers large, solitary or rarely spicate; fruit with very numerous minute seeds:
Flowers bisexual; anthers crowded into 3 or 4 sessile masses; flowers never very large *Hydnoraceae*
Flowers unisexual; anthers in 1–3 whorls around the column of the gynoecium; flowers sometimes exceptionally large *Cytinaceae*
Ovules 3 from the apex of a central placenta; flowers dioceious *Misodendraceae*
Ovule solitary, pendulous:
Flowers densely crowded into androgynous or unisexual inflorescences; fruits nut-like, 1-seeded; anthers opening by slits *Balanophoraceae*
Flowers spicate, racemose, or capitate; anthers opening by valves *Lauraceae*

KEY TO THE ARTIFICIAL GROUPS OF MONOCOTYLEDONS

Ovary completely superior:
Perianth present, or if small or very reduced or modified, or absent, then flowers not accompanied by glumaceous bracts or bracteoles (glumes):
Carpels free or only slightly united at the base, or gynoecium reduced to 1 carpel with 1 stigma; mostly aquatic, sometimes maritime plants
Group 1 (p. 102)
Carpels more or less completely united, with usually more than 1 stigma; rarely aquatic:
Perianth composed of separate calyx and corolla, the former often green, the latter usually petaloid, sometimes both series dry and hyaline but never united into one tube Group 2 (p. 103)
Perianth composed of similar or subsimilar segments in two or one series, usually very conspicuous and petaloid, if united then connate in the lower part into a single tube, sometimes (when inflorescence a spadix) very small and inconspicuous Group 3 (p. 105)
Perianth sepaloid or dry and glumaceous, usually very small or nothing; flowers mostly small and inconspicuous, and arranged in spadices or panicles and often subtended by large spathaceous bracts; or plants grass-like or sedge-like with very small flowers Group 4 (p. 107)
Perianth absent or represented by 'hypogynous setae' or 'scales' or lodicules, with the flowers minute, arranged in spikelets and in the axils of scaly bracts (glumes) Group 5 (p. 108)
Ovary semi-inferior Group 6 (p. 108)
Ovary completely inferior:
Perianth composed of separate calyx and corolla, remaining in two distinct series, the calyx often green or different from the inner petaloid series
Group 7 (p. 109)
Perianth-segments more or less all alike and usually petaloid, mostly 6, sometimes 3, free or often united at the base into a single tube
Group 8 (p. 110)

Ovary completely superior; perianth present, or if small or very reduced or modified or absent, then flowers not accompanied by glumaceous bracts or bracteoles (glumes); carpels free or only 1 *carpel with* 1 *stigma.*

Flowers bracteate:
Ovules spread all over the inner surface of the carpels or intruding septa; carpels dehiscent; flowers often in umbels or solitary, usually involucrate with spathaceous bracts *Butomaceae*
Ovules inserted on a placenta or at the base or apex of the carpels;
Herbs with green leaves, not saprophytic; mostly aquatic or semi-aquatic:
Leaves not ligulate, often pellucid-punctate or lined; carpels usually numerous (more than 3), in an irregular bunch or more rarely in a whorl; petals often clawed, usually very different from the sepals, the latter mostly green *Alismataceae*
Leaves ligulate, subterete; carpels 3(–6), divaricate in fruit; perianth-segments subsimilar; stamens 6; anthers basifixed, extrorse; stigmas sessile *Scheuchzeriaceae*
Saprophytic herbs with very reduced scale-like colourless leaves:
Carpels 3, divaricate in fruit; perianth imbricate; flowers bisexual *Petrosaviaceae*
Carpels numerous, crowded; perianth valvate; flowers unisexual or polygamous *Triuridaceae*
Tall palms; leaves plicate in bud; flowers in panicles with spathaceous bracts *Arecaceae* (*Palmae*)
Flowers without bracts (sometimes the single perianth-segment somewhat bract-like):
Styles heteromorphous; flowers unisexual and bisexual, in the latter the style absent and the flowers in spikes *Lilaeaceae*
Styles homorphous or absent:
Terrestrial herbs with spike-like racemes or spikes of small flowers; carpels 6 or 4 *Juncaginaceae*
Freshwater aquatics:
Flowers in racemes clusters or spikes:
Flowers usually arranged on one side of the inflorescence; stamens 6 or more, on fairly long filaments; perianth-segments 3–1 or absent, sometimes bract-like *Aponogetonaceae*
Flowers all round the axis of the inflorescence:
Flowers bisexual; stamens 4, inserted on the claws of the perianth-segments; anthers subsessile; carpels 4 *Potamogetonaceae*
Flowers unisexual, in very dense cylindrical spikes *Typhaceae*
Flowers in globose clusters *Sparganiaceae*
Flowers axillary, solitary or in small cymes; stamens often solitary:
Carpels 2 or more; ovule pendulous *Zannichelliaceae*
Carpel solitary; ovule basal *Najadaceae*
Flowers minute, 1–3 in a minute membranous spathe; floating plants, with

minute cellular green frond-like plant body *Lemnaceae*

Marine or salt marsh plants:

Flowers spicate:

Flowers bisexual, all round the axis:

Carpels 4 or more *Ruppiaceae*

Carpel solitary *Posidoniaceae*

Flowers monoecious or dioecious; flowers on one side of a flattened axis *Zosteraceae*

Flowers axillary or cymose *Zannichelliaceae*

Group 2. *Ovary completely superior, syncarpous; perianth of separate calyx and corolla, the former often green, the latter usually petaloid, sometimes both series dry and hyaline but not united into one tube.*

Branches modified into cladodes and bearing the flowers on their surface or margins; leaves reduced to scales *Ruscaceae*

Branches and leaves not as above:

Flowers usually in cymes, panicles, racemes, or spikes or rarely solitary, but not capitate and without an involucre of bracts, sometimes enclosed by large leafy boat-shaped bracts or coloured leaves; leaf-sheath usually closed; seeds often with a distinct pore-like 'stopper' (embryostega):

Leaves plicate in bud; usually tall palms with large panicles of small flowers with large spathaceous bracts *Arecaceae* (*Palmae*)

Leaves not plicate and other characters above not present:

Ovary 3-2-locular, with the ovules on axile or basal placentas; leaf-sheaths closed:

Flowers without bracts; small herbs with spike-like racemes of small flowers *Juncaginaceae*

Flowers with bracts:
Perianth of (usually green) calyx and petaloid corolla:
Ovules few to 1; leaves never cirrhose at the tip; fruit usually a capsule; seeds with a small 'stopper' (embryostega):
Glandular hairs absent; flowers cymose or solitary *Commelinaceae*
Glandular hairs present; flowers spicate *Cartonemataceae*
Ovules numerous; fruit usually fleshy and indehiscent; anthers often versatile; seeds without a 'stopper' but often appendaged *Bromeliaceae*
Perianth dry or somewhat petaloid, small; ovule solitary; stems erect or climbing:
Fruit indehiscent; leaves often cirrhose at the tips *Flagellariaceae*
Fruit dehiscent; leaves never cirrhose at the tip *Xanthorrhoeaceae*
Ovary 1-locular, with 3 parietal placentas:
Herbs:
Leaves very narrow, crowded, linear or thread-like, not sheathing at the base, bidentate at the apex; stamens 3; style simple *Mayacaceae*
Leaves broad and green, opposite or whorled; stamens 6; style-branches 3 *Trilliaceae*
Woody climbers or shrubs *Philesiaceae*
Flowers in heads and very small, often surrounded by two or more involucral bracts, or unilaterally spicate within a spathe; perianth often more or less hyaline:
Flowers bisexual; ovules parietal or ascending from the base (to p. 104):
Stamens 6, without staminodes; anthers opening by a pore; sepals equal; involucral bracts (when present) foliaceous and often 2, or flowers unilaterally spicate and within a folded spathe *Rapateaceae*
Stamens 3, often with 3 staminodes; anthers opening by slits; sepals unequal, one larger than the others and forming a hood over the corolla *Xyridaceae*
Flowers unisexual, monoecious, crowded into small heads; inner perianth-segments often united; ovule solitary, pendulous; perennials or rarely annuals *Eriocaulaceae*

Group 3. *Ovary completely (mostly) superior, syncarpous; perianth of similar or subsimilar segments in two or one series, usually very conspicuous and petaloid, if united then connate in the lower part into a single tube, sometimes (when inflorescence a spadix) very small and inconspicuous.*

Flowers arranged in a scapose umbel subtended by more or less membranous spathaceous bracts; rootstock usually a bulb; leaves radical and mostly linear *Amaryllidaceae*

Flowers not in umbels, or if rarely subumbellate or rarely in heads then bracts not spathaceous:

Flowers in a spadix subtended by or enclosed in a spathe, very small and inconspicuous and often unisexual *Araceae*

Flowers not in a spadix and without a spathe:

Branches modified and leaf-like (cladodes) and bearing the flowers on their margins or surface, the true leaves reduced to scales; woody plants; filaments connate into a column; fruit a berry *Ruscaceae*

Branches not modified or if so (*Asparagus*) then not bearing the flowers:

Aquatic herbs; inflorescence subtended by a spathe-like leaf-sheath; seeds ribbed; floral bracts absent or very small; stems or branches bearing only 1 leaf *Pontederiaceae*

Terrestrial or marsh plants; inflorescence not subtended by a spathe-like leaf-sheath; floral bracts rarely absent:

Stamens more than 1; flowers actinomorphic ('regular') (to p. 106):

Anthers usually 2-locular; flowers mostly bisexual; leaves mostly with parallel nerves and veins (to p. 106):

Leaves alternate, or the lower alternate and the upper in a pseudowhorl at the top below the flowers (to p. 106):

Woody climbers or shrubs with reticulate leaves:

Stamens 6:

Ovules numerous *Philesiaceae*

Ovule solitary in each loculus *Xanthorrhoeaceae*

Stamens 4; ovules 2 to many *Roxburghiaceae*

Herbs or herbaceous climbers, usually with parallel nerved or veined leaves; or if woody or tree-like then the leaves in dense tufts at the base or apex of the stems:

Ovary completely superior; anthers opening by slits or very rarely by pores; rootstock a rhizome or bulb, rarely a corm:

Ovary composed of 6 carpels; flowers without bracts; inner perianth-segments inserted higher than the outer *Juncaginaceae*

Ovary composed or 3 carpels; flowers usually with bracts; fruit a capsule or berry:

Perianth not dry and not very small:

Plants not or only slightly xerophytic; leaves not fibrous (to p. 106):

Style usually divided *Liliaceae*

Style undivided *Haemodoraceae*

Plants usually very xerophytic; leaves mostly fibrous, in a dense tuft either at the base or apex of the stem; flowers mostly in large panicles *Agavaceae*

Perianth dry and small; plants usually markedly xerophytic; style simple *Xanthorrhoeaceae*

Ovary composed of 3 carpels; flowers mostly without bracts; fruit drupaceous *Flagellariaceae*

Ovary partly inferior; anthers opening by pores; rootstock a corm or flattened tuber *Tecophilaeaceae*

Leaves opposite, or all whorled in the middle or at the top of the stem and then subtending the flower or flowers; flowers terminal, solitary or several and subumbellate or paniculate:

Small herbs with soft herbaceous leaves and few flowers *Trilliaceae*

Large and often woody plants with fibrous leaves and usually large panicles of flowers *Agavaceae*

Anthers 1-locular by confluence of the loculi; flowers small, mostly dioecious; stem climbing or straggling, often prickly; leaves 3–5-nerved, with reticulate venation *Smilacaceae*

Stamen 1, at the base of the anticous segment; flowers very zygomorphic ('irregular') *Philydraceae*

Group 4. *Ovary completely superior, syncarpous; perianth sepaloid or dry and glumaceous, usually very small and inconspicuous, or rarely absent; flowers often arranged in spadices or panicles or heads, usually subtended by large spathaceous or leafy bracts; or plants grass-like or sedge-like.*

Leaves plicate in bud, with strong parallel nerves, often pinnately or flabellately divided or nerved:
- **Perianth**-segments 6, in two distinct series; mostly tall plants with simple or little-branched stems; flowers small, in panicles or spikes *Arecaceae* (*Palmae*)
- **Perianth**-segments 4 or many, or absent; dwarf plants with often deeply bilobed leaves; flowers in dense spikes *Cyclanthaceae*

Leaves not plicate in bud:
- **Inflorescence** without spathaceous but sometimes with glumaceous leaf-like bracts; leaves linear, often tufted at the base of the stem:
  - **Anthers** dorsifixed, versatile; style simple or nearly so *Xanthorrhoeaceae*
  - **Anthers** basifixed, erect:
    - **Flowers** bisexual:
      - **Perianth**-segments, when present, in whorls:
        - **Stamens** 3 or more; ovules numerous or few; leaf-blade usually well developed; perianth present; ovules, when solitary, erect *Juncaceae*
        - **Stamens** 1 or rarely 2; leaves small; perianth absent; ovule solitary, pendulous *Centrolepidaceae*
      - **Perianth**-segments irregularly arranged; flowers in globose or oblong heads subtended by long foliaceous bracts *Thurniaceae*
    - **Flowers** dioecious; leaves mostly reduced to sheaths; mainly Southern Hemisphere *Restionaceae*
- **Inflorescence** subtended by a spathaceous bract, or with spathaceous bracts:
  - **Flowers** bisexual, or if unisexual then monoecious, though sometimes separated on the spadix and often accompanied by neuter flowers:
    - **Terrestrial** or very rarely (*Pistia*) aquatic and then with well developed leaves *Araceae*
    - **Aquatic** plants, floating; plant-body minute, not differentiated into stem and leaves; flowers very minute, perianth absent *Lemnaceae*
  - **Flowers** unisexual, dioecious or very rarely monoecious; floral bracts absent:
    - **Herbaceous**, aquatic plants; leaves without prickles; flowers in dense cylindrical spikes *Typhaceae*
    - **Shrubs** or trees; margins of the usually spirally coiled leaves often prickly; aerial roots often present *Pandanaceae*

Group 5. *Ovary superior; perianth absent or represented by 'hypogynous' setae or scales or lodicules; flowers minute, generally in spikelets and in the axils of scaly bracts.*

Fruit dehiscent; small herbs mostly of the Southern Hemisphere; ovule pendulous; stamens 1 or rarely 2 *Centrolepidaceae*
Fruit indehiscent; ovule erect or ascending:
Flowers in the axil of a single bract and collected into spikelets, the latter variously arranged, from solitary to umbellate, paniculate, or capitate; anthers basifixed; leaves usually with closed sheaths; stems mostly solid and triquetrous; embryo free from the pericarp *Cyperaceae*
Flowers enclosed by a bract and bracteole (lemma and palea), arranged in spikelets; anthers mostly dorsifixed and versatile; leaves usually with open sheaths, the latter mostly tipped with a ligule; stems mostly with hollow internodes and usually terete; embryo usually adnate to the pericarp *Poaceae* (*Gramineae*)

Group 6. *Ovary semi-inferior.*

Perianth-segments in two distinct series, the outer calyx-like, the inner petaloid; leaves mostly linear and spinulose on the margin *Bromeliaceae*
Perianth-segments similar to subsimilar, sometimes united into one tube:
Anthers opening by pores; rootstock a corm:
Perianth persistent, not circumscissile at the base *Haemodoraceae*
Perianth deciduous, circumscissile at the base *Tecophilaeaceae*
Anthers opening by slits lengthwise:
Stamens and perianth-segments 6:
Filaments free from one another:
Ovules 2; fruit bursting before maturity and exposing the unripe seeds *Liliaceae* (*Ophiopogoneae*)
Ovules numerous *Haemodoraceae*
Filaments connate into an annulus *Liliaceae* (*Peliosantheae*)
Stamens and perianth-segments 4; rhizome creeping; leaves with close transverse veins *Roxburghiaceae*

Group 7. *Ovary completely inferior, syncarpous; perianth of separate calyx and corolla, remaining in two distinct series, the calyx often green or different from the inner petaloid series.*

Inner perianth actinomorphic ('regular'):
Stamens 3 or more; no petaloid staminodes:
Aquatic; ovules spread all over the inner surface of the carpels or rarely superposed in 2 series; flowers mostly unisexual, solitary or the male several within 2 folded or 1 bifid bracts; outer perianth valvate or induplicate; no endosperm *Hydrocharitaceae*
Terrestiral or epiphytic; ovules confined to placentas; flowers bisexual or very rarely by abortion dioecious:
Calyx actinomorphic ('regular'), lobes imbricate; endosperm copious:
Stamens 6; bracts often coloured *Bromeliaceae*
Stamens 3; bracts usually membranous *Iridaceae*
Calyx tubular, soon split down one side, 3–5-dentate at the apex; bracts large and spathaceous; flowers unisexual *Musaceae*
Stamen 1 fertile, the remainder transformed into petaloid staminodes often more conspicuous than the inner perianth:
Anthers 2-locular; sepals united into a sometimes spathaceous tube; staminodes adnate to the base of the inner perianth; ovules numerous; embryo straight, central; mostly Old World tropics *Zingiberaceae*
Anthers 1-locular; sepals free or at most connivent:
Ovules numerous in each loculus; embryo straight, central; tropics generally *Cannaceae*
Ovules solitary in each loculus; embryo much curved; mostly New World tropics *Marantaceae*
Inner perianth zygomorphic ('irregular');
Stamens 6 or 5:
Leaves and bracts spirally arranged; calyx more or less spathaceous; fruit indehiscent *Musaceae*
Leaves and bracts distichous:
Sepals 3, free *Strelitziaceae*
Sepals united into a tube *Lowiaceae*
Stamens 3; ovary not twisted; pollen granular; seeds with endosperm *Iridaceae*
Stamens 2 or 1, inserted on a prolongation of the axis (the column), with often the pollen agglutinated into masses; ovary often spirally twisted; seeds very numerous and minute, without endosperm; one of the petals (lip) different from the others *Orchidaceae*

Group 8. *Ovary completely inferior, syncarpous; perianth-segments usually petaloid, mostly* 6, *sometimes* 3, *free or often united at the base into a single tube.*

Ovules spread all over the inner walls of the carpels or on the instrusive septa; flowers borne in spathaceous bracts; no endosperm *Hydrocharitaceae*
Ovules borne on the placentas or at the base or apex of the ovary:
Inflorescence scapose, umbellate, subtended by an involucre of one or more spathaceous bracts, sometimes the flower solitary and subtended by one or more spathaceous bracts:
Perianth actinomorphic; ovary not twisted; stamens 6; seeds with endosperm *Amaryllidaceae*
Perianth zygomorphic ('irregular'); ovary usually twisted; stamens 2 or 1; seeds without endosperm *Orchidaceae*
Inflorescence not as above, or if appearing to be umbellate through crowding then not subtended by an involucre of spathaceous bracts (sometimes the leaves are more or less whorled below the flowers):
Small saprophytic herbs, usually with much reduced scale-like, colourless (rarely green) leaves:
Stamens 6 or 3; ovary not twisted:
Perianth actinomorphic ('regular'):
Perianth-tube cylindrical, shortly lobed, lobes not appendaged; stamens 3; ovary and fruit winged *Burmanniaceae*
Perianth inflated or campanulate, with filiform or appendaged lobes; stamens usually 6 *Thismiaceae*
Perianth zygomorphic, one of the outer segments large and ovate-cordate, the remainder linear stamens 6 *Corsiaceae*
Stamen 1; perianth zygomorphic ovary usually twisted *Orchidaceae*
Not saprophytic:
Stamens 6 or more (to p. 111):
Fruit a capsule (to p. 111):
Stems leafy (sometimes only 1 leaf); leaves scattered on the stem or often the leaves all crowded at the apex of the shoots (to p. 111):
Flowers bisexual, usually very showy (to p. 111):
Seeds not winged or very slightly so; capsule not very elongated:
Leaves not all crowded at the apex of the stem or branches; stems not woody:
Perianth-segments free, sometimes somewhat unequal either in shape or colour, often mottled or spotted *Alstroemeriaceae*
Perianth-segments united at the base, subequal; leaf solitary on each shoot *Trichopodaceae*
Leaves all crowded at the apex of the woody stem, fibrous, the basal sheaths persistent and forming a dense covering *Velloziaceae*
Seeds winged; capsule much elongated; perianth-segments united into a tube; leaves with numerous nerves and cross-nerves *Stenomeridaceae*

Flowers unisexual, usually very small and inconspicuous; seeds mostly winged; usually climbers *Dioscoreaceae*

Stems with a tuft of leaves only at the base, or leaves all radical, or stem leaves much reduced and smaller than the basal leaves:

Perianth usually very woolly-tomentose outside, the hairs often branched; embryo marginal, not wholly enclosed by the endosperm *Haemodoraceae*

Perianth usually glabrous, or if hairy then hairs not branched; embryo enclosed by the endosperm; leaves mostly fibrous:

Inflorescence scapose, radical; flowers solitary to spicate or racemose; perianth without a tube (but sometimes the ovary long-beaked and resembling a narrow tube) *Hypoxidaceae*

Inflorescence terminal, long-racemose, spicate or paniculate, sometimes very large; perianth usually with a distinct tube *Agavaceae*

Fruit a berry or indehiscent:

Herbs (not climbing):

Leaves entire:

Leaves linear, with parallel nerves *Hypoxidaceae*

Leaves more or less ovate, with pinnate nerves *Trichopodaceae*

Leaves often much divided; flowers subumbellate *Taccaceae*

Climbers; leaves often broad and with reticulate venation, sometimes deeply divided:

Ovary 3-locular; herbaceous climbers, rarely somewhat woody; ovules 2 in each loculus, superposed *Dioscoreaceae*

Ovary 1-locular, with parietal placentas and numerous ovules *Petermanniaceae*

Stamens 3; perianth rarely zygomorphic; pollen granular:

Climbers, with usually broad sometimes compound leaves and small inconspicuous flowers *Dioscoreaceae*

Herbs, with rhizomes, corms, or rarely bulbs:

Filaments not adnate to the style:

Style lobed in the upper part; indumentum absent or if present not of branches hairs; embryo not marginal *Iridaceae*

Style entire or nearly so; indumentum often of branches hairs; embryo maringal *Haemodoraceae*

Filaments connate in the lower part and with the style; flowers spicate or racemose; style entire, with 3 minute stigmas *Apostasiaceae*

Stamen 1; pollen often agglutinated; ovary often twisted; perianth strongly zygomorphic *Orchidaceae*

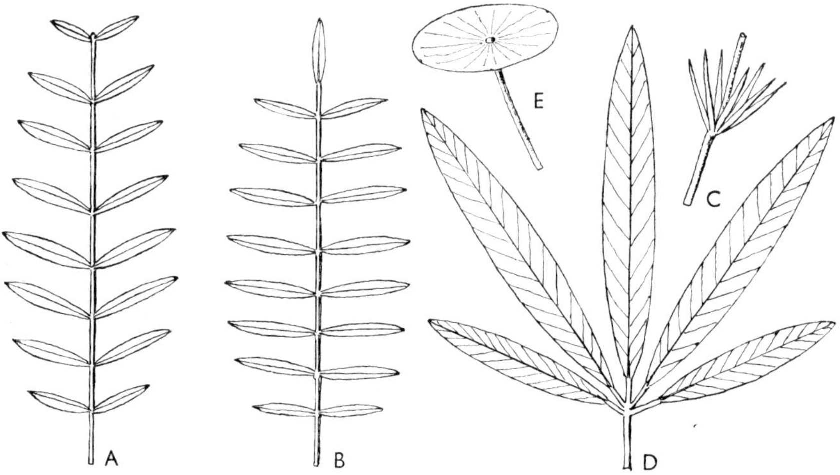

FIG. 1. Leaves: A, paripinnate; B, inparipinnate; C, verticillate; D, digitately compound; E, peltate.

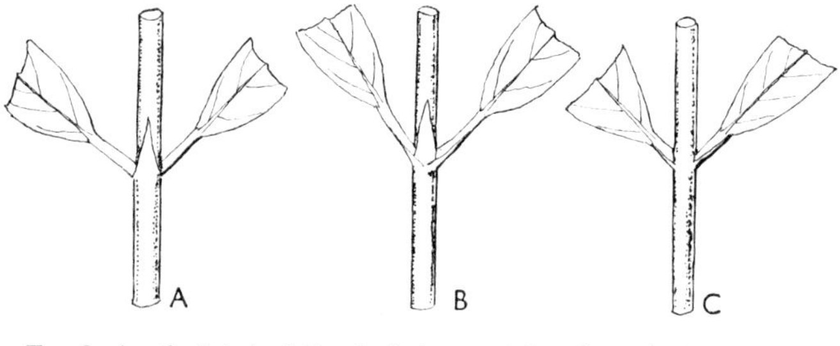

FIG. 2. A, stipulate leaf. B, stipule intrapetiolar. C, exstipulate leaves.

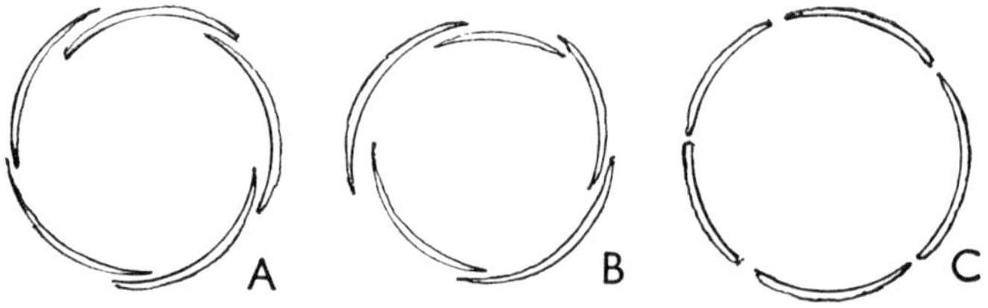

FIG. 3. Types of aestivation: A, contorted; B, imbricate; C, valvate

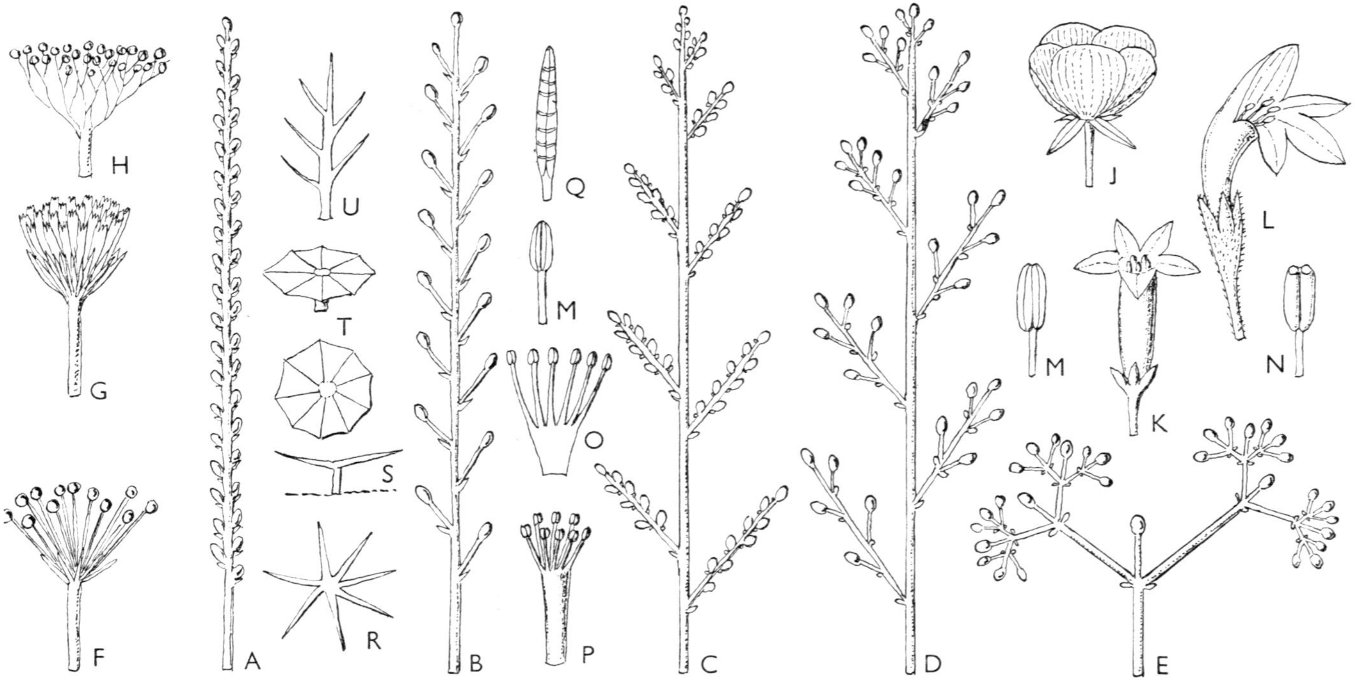

FIG. 4. Various kinds of inflorescences, flowers, etc. A, spike; B, raceme; C, panicle of spikes; D, panicle of racemes; E, dichotomous cyme; F, umbel; G, capitulum; H, corymb; J, actinomorphic flower, with free petals; K, sympetalous corolla; L, zygomorphic flower; M, stamens; N, anther opening by terminal pores; O, bundle (phalange) of stamens; P, monadelphous stamens; Q, transversely locellate anther; R, stellate hair; S, medifixed hair; T, scaly hair (lepidote); U, dendriform hair.

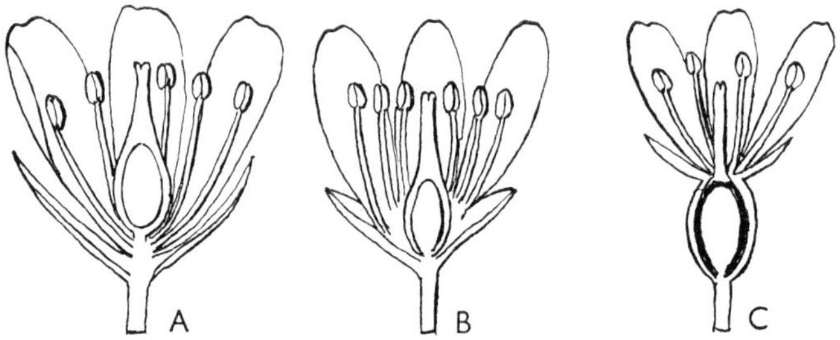

FIG. 5. A, hypogynous flower. B, perigynous flower. C, epigynous flower.

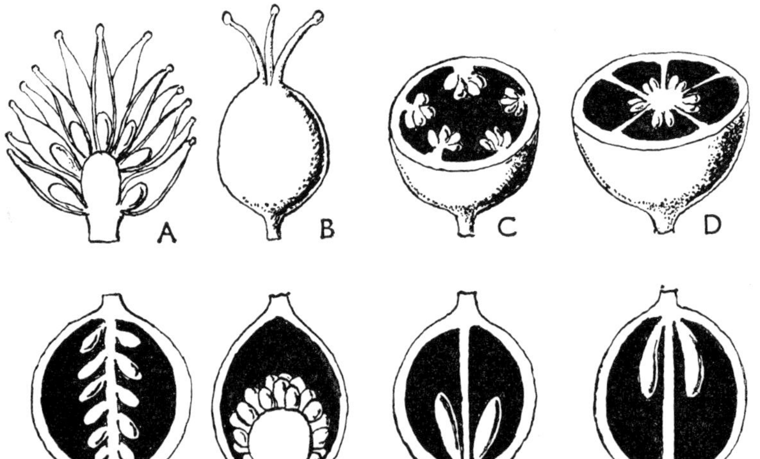

FIG. 6. Gynoecium: A, apocarpous; B, syncarpous; C, parietal placentation; D, axile placentation; E, free central placentation; F, free basal placentation; G, basal erect ovules; H, apical pendulous ovules.

GLOSSARY

abaxial, away from the axis
accumbent, lying against
achene, small dry indehiscent fruit containing a single seed (as in *Ranunculus*)
actinomorphic ('regular'), capable of vertical division into equal halves through two or more planes.
adaxial, next to the axis
adnate, more or less united to a member of another series (as petals in *Malvaceae*)
aestivation, arrangement of sepals or petals in bud
anemophilous, wind-pollinated
annual, lasting for one or part of one year
apocarpous, gynoecium of separate carpels (as in *Ranunculaceae*)
aril, appendage arising from the funicle and sometimes covering the seed
articulate, jointed
axile, attached to the central axis; also used for flowers in the axil of the leaf

baccate, applied to a succulent fruit with seeds immersed in the pulp
biennial, lasting for two seasons
bilabiate, 2-lipped (as in *Lamiaceae*)
bisexual, with two sexes in the same flower or inflorescence

calyx, outer whorl of a flower, usually green or scarious
capitate, head-like
capsule, dry fruit (composed of more than one carpel) which opens to release the seeds
carpel, a separate or united member of the gynoecium
caruncle, a wart-like protuberance of a seed (as in many *Euphorbiaceae*)
catkin, a close bracteate often pendulous spike
circinate, inwardly coiled upon itself like a watch-spring
circumscissile, opening transversely all around
cirrhose, provided with tendrils
cleistogamous, fertilization within the unopened flower
compound, the opposite of simple; composed of several separate parts (as in the leaf of *Aesculus*)
conduplicate, folded face to face
connate, when similar parts are more or less united (as in a sympetalous corolla)
contorted, folded in bud, such as sepals or petals which overlap on one side only
corolla, the inner whorl of the flower, often coloured and attractive
corona, a ring of appendages (sometimes united) on the inside of the petals or corolla (as in *Passiflora*)
corymb, a more or less flat-topped inflorescence (as in many *Asteraceae*)
cotyledon, seed-leaf
cyclic, with floral parts in whorls
cyme, a broad divaricately branched inflorescence

deciduous, falling off (as many leaves)
dehiscent, opening spontaneously
diadelphous, in two bundles (as in *Fabaceae*), often with one stamen separate from the others
dicotyledon, a plant with two seed-leaves
didymous, in pairs, as the divergent lobes of anthers or fruits
digitate, diverging from the same point (as the leaflets of *Aesculus*)
dimorphic, of two forms
dioecious, with male flowers on one individual, females on another
discoid, like a disk; also applied to *Asteraceae* without ray-flowers
disk, either an enlargement of the torus or formed of the union of the nectaries within the calyx, petals or stamens
dissepiments, partitions of an ovary or fruit

embryo, rudimentary plant still enclosed in the seed
endosperm, the nutritive reserve food material within some seeds
entomophilous, insect-pollinated
epigynous, inserted above the gynoecium
epipetalous, on the petals or corolla
epiphyte, plant growing or supported on another
exstipulate, without stipules
extrorse, facing outwards (applied to anthers facing away from the centre of the flower)

follicle, fruit of a single free carpel opening by one suture
free basal placentation, ovules arranged on a basal placenta in a 1-locular ovary (as in *Primula*)

free central placentation, ovules on a central placenta reaching from the bottom to the top of a 1-locular ovary (as in many *Caryophyllaceae*)
funicle, cord or thread connecting ovule or seed to the placenta

gamopetalous (=*sympetalous, monopetalous*) a corolla formed of more or less united petals (as in *Rhododendron*)
gamosepalous, sepals more or less united
glandular, clothed with glands
glochidiate, furnished with barbed bristles
gynoecium (pistil), the female part of the flower (in the middle), consisting of free or united carpels, with style and or stigma
gynobasic, a style which arises from near the base of the carpels or lobes of the ovary
gynophore, a stalk supporting the ovary within the stamens

herb, plant not woody
hilum, scar of attachment on a seed
hypogynous, parts of the flower inserted below the gynoecium (pistil)

imbricate, overlapping; applied to sepals or petals with one or more wholly outside or inside and some half out and half in
incumbent, lying on or against (as the cotyledons in relation to the radicle in many *Brassicaceae*)
indehiscent, not opening
indumentum, hairy covering
induplicate, folded inwards
indusiate, like a small cup
inferior, below, as the ovary below the other parts of the flower
interpetiolar, between the petioles
intrapetiolar, within the petiole
introrse, facing inwards (applied to anthers)
irregular, see *zygomorphic*

legume, pod formed of one carpel opening by one or two sutures (as in *Fabaceae*)
lepidote, clothed with scales
locellate, subdivided into smaller loculi
locular, chambered ('celled'), of an ovary or anther
loculi, chambers ('cells') of an ovary or anther
loculicidal, opening into the loculi

marcescent, withering without falling
monadelphous, in one bundle (as stamens in *Malvaceae*)
monochlamydeous, a flower with only one whorl of the perianth (usually a calyx)
monocotyledon, a plant with only one seed-leaf
monoecious, unisexual flowers on the same plant
monopetalous, petals more or less united, or one petal
muticous, without a point

ochreate, sheathing, as the stipules in *Polygonaceae*
opposite, opposed or inserted at the same level
ovary, the lower part of the gynoecium containing the ovules
ovule, the potential seed

panicle, a branched spike or raceme
papillous, with short protuberances
parietal placentation, ovules attached to the walls of a 1-locular syncarpous ovary
pellucid, translucent
pendulous, handing down
perennial, lasting for more than two seasons
perianth, the floral envelope or envelopes
perigynous, placed around the ovary but free from it
phalange, a bundle of stamens (as in *Hypericum*)
pistil, see *gynoecium*
placenta, the part of the ovary to which the ovules are attached.
plicate, folded like a fan
pollen, the powdery contents of an anther (the male element of the flower)
polygamous, flowers male, female and bisexual on the same plant
polypetalous, petals free from each other
protandrous, anthers ripening before the stigmas
protogynous, stigmas receptive before the anthers open
punctate, dotted

racemose, inflorescence with stalked flowers on an unbranched axis
radical, arising from the root or base of a plant
radicle, root of embryonic plant
receptacle, the part of the floral axis bearing the sepals, petals, stamens and carpels; also applied to the axis of the head of a Composite flower
regular, see *actinomorphic*
ruminate, endosperm into which the inner layer of the testa protrudes

septate, divided by partitions
septicidal, a fruit which opens by the septa between the loculi
simple, the opposite of compound

spike, inflorescence with sessile flowers on a single elongated axis or rhachis
spur, a hollow horn-like extension of a part of the flower
staminode, an abortive, sometimes much modified or reduced stamen
stellate, star-like
stigma, the receptive part of the style
stipe, the stalk of a carpel or gynoecium
stipitate, furnished with a stipe
stipel, stipule-like appendages to the leaflets of a compound leaf
stipule, basal appendage to a leaf or petiole
stoma, (pl. *stomata*), the breathing apparatus in the epidermis of leaves
style, the terminal part of the pistil bearing the stigma
superior, above, as an ovary above the other parts of the flower
superposed, placed one above another
sympetalous, a corolla with the petals partially or wholly united
syncarpous, gynoecium composed of united carpels

testa, coat of a seed
tetradynamous, four long and two short (as stamens in *Brassicaceae*)
torus, see receptacle

umbel, an inflorescence the branches of which spring from the same point (characteristic of *Apiaceae* (*Umbelliferae*))
unisexual, of one sex

valvate, margins of sepals or petals not overlapping
valves, of anthers which open by flaps
vernation, folding of leaves in bud stage

zygomorphic ('irregular'), flowers which may be divided into equal halves in only one plane